T0252953

Textbooks in Telecommunication Engineering

Series Editor
Tarek S. El-Bawab, Professor and Dean,
School of Engineering
American University of Nigeria
Yola, Nigeria

Telecommunication and networks have evolved to embrace all aspects of our everyday life, including smart cities and infrastructures, healthcare, banking and businesses, manufacturing, space and aviation, meteorology and climate change, oceans and marine life, Internet of Things, defense, homeland security, education, research, social media, entertainment, and many others. Network applications and services continue to expand, virtually without limits. Therefore, specialized telecommunication and network engineering programs are recognized as a necessity to accelerate the pace of advancement in this field, and to prepare a new generation of engineers for imminent needs in our modern life. These programs need curricula, courses, labs, and textbooks of their own.

The IEEE Communications Society's Telecommunication Engineering Education (TEE) movement, led by Tarek S. El-Bawab- the editor of this Series, resulted in recognition of this field of engineering by the Accreditation Board for Engineering and Technology (ABET) on November 1, 2014. This Springer Series was launched to capitalizes on this milestone (with the original title "Textbooks in Telecommunication Engineering"). The Series goal is to produce high-quality textbooks to fulfill the education needs of telecommunication and network engineering, and to support the development of specialized undergraduate and graduate curricula in this regard. The Series also supports research in this field and helps prepare its scholars for global challenges that lay ahead. The Series have published innovative textbooks in areas of network science and engineering where textbooks have been rare. It is producing high-quality volumes featuring innovative presentation media, interactive content, and online resources for students and professors.

Book proposals are solicited in all topics of telecommunication and network engineering including, but not limited to: network architecture and protocols; traffic engineering; network design, dimensioning, modeling, measurements, and analytics; network management and softwarization; cybersecurity; synchronization and control; applications of artificial intelligence in telecommunications and networks; applications of data science in telecommunications and networks; network availability, reliability, protection, recovery and restoration; wireless communication systems; cellular technologies and networks (through 5G, 6G, and beyond); satellite and space communications and networks; optical communications and networks; heterogeneous networks; broadband access and free-space optical communications; MSO/cable networks; storage networks; optical interconnects; and data centers; social networks; transmission media and systems; switching and routing (from legacy to today's paradigms); network applications and services; telecom economics and business; telecom regulation and policies; standards and standardization; and laboratories.

Proposals of interest shall be for textbooks that can be used to develop university courses, either in full or in part. They should include recent advances in the field while capturing whatever fundamentals that are necessary for students to understand the bases of the topic and appreciate its evolution trends. Books in this series will provide high-quality illustrations, examples, end-of-chapters' problems/exercises and case studies.

For further information and to submit proposals, please contact the Series Editor, Dr. Tarek S. El-Bawab, telbawab@ieee.org; or Mary James, Executive Editor at Springer, mary.james@springer.com

M. Scott Kingsley

Cloud Technologies and Services

Theoretical Concepts and Practical Applications

 Springer

M. Scott Kingsley
Department of Electrical and Computer Engineering
Lyle School of Engineering
Southern Methodist University
Dallas, TX, USA

ISSN 2524-4345   ISSN 2524-4353   (electronic)
Textbooks in Telecommunication Engineering
ISBN 978-3-031-33668-3   ISBN 978-3-031-33669-0   (eBook)
https://doi.org/10.1007/978-3-031-33669-0

This Springer imprint is published by the registered company Springer Nature Switzerland AG
The registered company address is: Gewerbestrasse 11, 6330 Cham, Switzerland

Paper in this product is recyclable.

Acknowledgements

I want to thank all of the students who were Teaching Assistants and those who volunteered for advanced independent study that contributed in various ways to the completion of this book (names are in alphabetical order):

Aneek Bera
Vikas Beri
Pranali Chintam
Abilash Gangali
Neeraj Gusain
Vishakha Kamble
Ishan Kembavi
Vishal Kulkarni
Rohan Kulkarni
Heny Laisham
Ashish Mahajan
Dylan Martinez
Shubhaum Padwankar
Kalyani Patthi
Rakesh Pyata
Himanshu Sakat
Vidhi Sethi
Shantanu Thakulr
Sai Usa

A few students were a vital part of this effort and deserve special recognition:

Palak Ghandi
Prasad Bhavsar
Nay Zaw

A special thank you to Mrs. Connie E. Howell whose professional editing expertise was invaluable.

Most of all I want thank my wife Lynda for her patience and encouragement through some very difficult times. Without her this book would have never been completed.

Sincerely,
Dr. Scott Kingsley

Department of Electrical and Computer Engineering   M. Scott Kingsley
Lyle School of Engineering
Southern Methodist University
Dallas, TX, USA

Contents

Part II   Cloud Networking

Part I
Introduction to Cloud Computing

You are reading this book because you have investigated "the cloud" and are interested in the possibility of pursuing a career in it. I am sure you have many questions. What do I need to learn? Where can I get the training or education? How hard is it to learn? How much money can I make? These questions and more will be answered for you in the pages that follow. However, we need to start from the beginning.

Chapter 1 will walk you through the basic building blocks of cloud infrastructures. These building blocks will then be used to illustrate the most common cloud application, the three-tier website.

So now you've taken the first steps on your cloud journey!

Chapter 1
Cloud Computing Concepts

1.1 The "Cloud"

The concept of the "cloud" is not new. The Internet has historically always been referred to as such in terms of a network that automatically routes traffic.

Later, the term "cloud computing" evolved to describe the access to remotely located computing resources. Users connected to those computing resources across a network. Computing capability could be increased or decreased on demand to meet even very large computing needs. As a result, users paid only for the computing capacity they used, which can be much less expensive than building and maintaining an onsite data center.

As the cloud evolved, the term "cloud computing" became very limiting. It now includes not only computing resources but other capabilities as well. In general, cloud computing now includes the basic functions of compute, storage, database, and networking. Understanding cloud computing requires a basic knowledge of each of these functions.

Using cloud computing only requires the ability to access the Internet. If you use Microsoft Office 365 for email or order a book on Amazon, you are using cloud computing. Hundreds of other public and privately developed applications are available.

How you get connected to your email or other applications is of little concern to most people. However, networks are the fabric that weaves cloud computing functions together. Without fast, reliable, and highly available networks, in particular, the Internet, cloud computing would not be possible.

M. S. Kingsley, *Cloud Technologies and Services*, Textbooks in Telecommunication Engineering, https://doi.org/10.1007/978-3-031-33669-0_1

1.2   Cloud Computing Defined

The National Institute of Standards and Technology (NIST) is responsible for developing guidelines for Information Technology (IT) systems used by the US federal government. Their definition of cloud computing is the most widely accepted (Mell & Grance, 2011):

> Cloud computing is a model for enabling ubiquitous, convenient, on-demand network access to a shared pool of configurable computing resources (e.g., networks, servers, storage, applications, and services) that can be rapidly provisioned and released with minimal management effort or service provider interaction. This cloud model is composed of five essential characteristics, three service models, and four deployment models.

Details of the NIST cloud categories are discussed further below.

1.3   Cloud Essential Characteristics

NIST defines five essential characteristics, including on-demand self-service, broadband network access, resource pooling, rapid elasticity, and measured service.

1.3.1   On-Demand Self-Service

The user can enable compute, storage, network, and other functions automatically with little or no human interaction. In other words, cloud computing users do not have to install, configure, cable, and operate the accessed resources.

1.3.2   Broadband Network Access

A fast, efficient, and universally available network allows access to cloud resources from a wide variety of different devices such as workstations, laptops, tablets, or cell phones. Networks include the Internet and public wireline of wireless networks.

1.3.3   Resource Pooling

Resources of compute, storage, database, and networking are pooled and provide "multitenancy" where different groups of users can enable and share those resources securely without the need to know their physical location.

1.3.4   Rapid Elasticity

Resources are rapidly and automatically scalable, or increased and decreased, to meet greater or lesser demand for those resources.

1.3.5   Measured Service

A public cloud provider facility has thousands of IT and networking devices that have been installed, cabled, hardware configured, and software installed. They are available for near immediate use. As a result, they can offer "pay-as-you-go" service to their customers. In other words, the customer can quickly increase or decrease cloud resources and pay only for them during the time they are actually in use.

1.4   Cloud Deployment Models

NIST defines four deployment models: public, private, community, and hybrid clouds. Also introduced is the "multicloud."

1.4.1   Public Clouds

When talking about the "cloud," most people are thinking about a *public* cloud. A public cloud is provided by a company, for example Amazon Web Services, which allows customers to remotely access and use their physical resources for a fee.

Public cloud providers are usually very large data centers that can be massive in size. They routinely house hundreds to thousands of hardware devices with some growing to as large as one million servers and related hardware in one facility.

Large cloud providers also have many geographically separated data centers. When a customer uses the cloud provider's resources, most are backed up in at least one other data center. When a failure occurs, the demand on the failed resources is automatically shifted to the backup systems.

Advantages of a public cloud provider include the following:

- **Decreased capital investment.** There is no need to purchase, install, and maintain hardware or software resources at the customer's location.
- **Agility.** Cloud provider resources are ready to be enabled or disabled in minutes rather than days or weeks.
- **Economics.** Customers pay only for the resources they use when they use them.
- **Redundancy.** Cloud resources are configured for backup operations. Recovery from failures is rapid and automatic, often at no additional cost to the customer.

However, there are several disadvantages to a public cloud provider:

- **Reduced resource control.** Using a public cloud provider results in a significant loss of control over compute and other resources. The degree of loss of control depends on the cloud service used.
- **Reduced control over data.** Data stored on a public cloud provider's resources is now stored and managed by the provider. The consumer has little control over the location or management of their data.
- **Security.** The customer must trust the public cloud provider to secure their cloud infrastructure while still accepting considerable security responsibility for applications placed on it.
- **Contract complexity.** Contracts enforced by public cloud providers are complex and still heavily favor the cloud provider. Clauses that address disputes, such as dissatisfaction with cloud services delivered even if proven, can result in large exit expenses and difficulty in moving to another provider.

There are many public cloud providers. However, the public cloud provider industry requires the investment of massive financial resources with no guarantee of success. Many large companies have entered the market only to withdraw later due to intense competition.

Attempting to rank public cloud providers' positions in the market is difficult since the industry is so volatile. What is accurate today may not be in the long term (Table 1.1).

Currently the three leading public cloud providers are Amazon Web Services, Microsoft Azure, and the Google Cloud Platform, respectively. However, their ranking may fluctuate in the future.

1.4.2   Private Clouds

When discussing the "cloud," it is generally referring to the public cloud providers. However, an enterprise can build their own private cloud for their internal use. They purchase, install, configure, operate, and maintain their own equipment in their own facilities including providing electricity and climate control systems.

Table 1.1 Top global public cloud providers (Gartner, 2023)

1.	Amazon Web Services
2.	Microsoft Azure
3.	Google Cloud Platform
4.	IBM Cloud
5.	Oracle Cloud Infrastructure (Gen2)
6.	VMware Cloud on AWS
7.	Alibaba Cloud
8.	Huawei Cloud
9.	Digital Ocean
10.	Zstack Enterprise

Advantages of a private cloud infrastructure include the following:

- **Control:** The user maintains complete control of the cloud infrastructure.
- **Customization:** The enterprise can customize their cloud infrastructure as needed.
- **Compliance:** Many industries have strict privacy and other compliance requirements that cannot be met using a public cloud provider.
- **Security:** Access to a private cloud can be confined to secured connections within the company.

Although significant cost advantages can be realized by implementing a private cloud over legacy IT infrastructure, there are disadvantages compared to using a public cloud provider.

- **Redundancy.** Public cloud providers provide redundant systems across geographically separated areas by default. Although a private cloud can implement resource redundancy, the cost of doing so can be exorbitant. Few enterprises can afford the level of redundancy, in particular geographic redundancy, that a public cloud provider offers. Therefore, the private cloud only provides limited redundancy to protect the most critical applications, which leaves others more vulnerable to failures.
- **Employee skillsets.** The enterprise has to maintain a competent workforce to manage and maintain the private cloud infrastructure. Employees with cloud technology skillsets have to be trained or hired. Time will be required for existing employees to broaden their skill portfolios in order to be relevant in the new environment.
- **Security.** In a private cloud, the user has full responsibility for all security. Highly skilled and expensive people will be needed to provide complex and extensive security measures. Public cloud providers have extensive expertise and resources focused on the security of their cloud infrastructure.
- **Compliance.** Regulatory compliance can often be difficult for a private cloud. Public cloud providers are often able to provide security and privacy compliant services required by HIPPA, Sarbanes-Oxley, FERPA, and others.
- **Opportunity cost.** Although a private cloud may offer cost benefits over a legacy IT infrastructure, the "opportunity," or possible, cost savings that can be realized from using a public cloud provider can be significant. Private clouds often cannot compare with the economies of scale offered by public cloud providers.
- **Cloud management.** Systems used to manage a legacy IT infrastructure are often not useable in a private cloud. There are many options for cloud management available at differing price points. OpenStack is an open source and therefore a free product capable of managing even a massive public or private cloud environment. Another popular product is from the private company VMware, but the cost can be prohibitive. Both require a high level of technical expertise to use.

1.4.3   Community Clouds

A community cloud intends to serve a diverse group of common users. A large enterprise may choose to implement a community cloud. However, they will probably do so by using a public cloud provider. Since the community cloud model is specialized, it will not be discussed further.

1.4.4   Hybrid Clouds

A hybrid cloud uses a combination of the private and public cloud models. Due to security concerns, the enterprise may want to keep sensitive data isolated to their on-premise private cloud resources while outsourcing their less sensitive applications, such as email, to a public cloud provider. A hybrid model could also be implemented that will allow the slow migration of their data and applications completely to a public cloud provider. The hybrid cloud is the most commonly implemented model for large enterprises.

1.4.5   Multicloud

Although it is not included in the standard NIST cloud computing definition, it is necessary to add one more model to our discussion.

As the cloud industry has evolved, almost half of all enterprises use multiple public cloud providers (Fleera, 2022). One reason is to avoid the risk of using just one provider. For example, it is common for an enterprise to use both Amazon Web Services (AWS) and Microsoft Azure or Google Cloud Platform (GCP). If one provider's strategy changes from that of the enterprise or their services are unsatisfactory, they have an alternative they can still rely on. Using two services can also benefit the enterprise during contract negotiations, including pricing.

Not all public cloud providers offer the same services. Large cloud providers may not offer specialty applications that create a niche market for smaller public cloud providers.

It is not uncommon for a large enterprise to use services from as many as a dozen different public cloud providers. This results in a new industry challenge on how to manage the enterprises infrastructure across numerous cloud providers.

1.5   Cloud Service Models

NIST defines three service models: Software-as-a-Service (SaaS), Platform-as-a-Service (PaaS), and Infrastructure-as-a-Service (IaaS) (Fig. 1.1). Not defined by NIST is the newer concept of "Anything-as-a-Service."

Fig. 1.1 Cloud service
models

1.5.1   Software-as-a-Service (SaaS)

In SaaS, the cloud provider allows access to software applications. The software developer will "host," or install, their application on a public cloud provider's remote facility. The consumers of the application are users who pay for the right to access and use the software. The consumer has little or no control over the configuration or operation of the software.

An example of SaaS is Microsoft Outlook 365. The consumer does not have to install and maintain their own on-premise email system. Instead, they pay Microsoft on a per-user basis for Microsoft to remotely host email services, which allows the consumers to send and receive emails across the network from the remote system.

1.5.2   Platform-as-a-Service (PaaS)

PaaS is intended primarily for application software developers to build software. Software developers need numerous tools to create software. These include operating systems, programming languages, as well as tools for software testing and integration. These tools are expensive and must, of course, run on costly hardware servers.

PaaS users have remote access to these tools on a per-user basis. Software developers no longer have to provide the hardware and software platforms to develop software. However, the user has no control over the underlying hardware infrastructure but has some control over the configuration and operation environment of the software tools.

1.5.3   Infrastructure-as-a-Service (IaaS)

IaaS allows remote access to IT resources located in a public cloud provider's facility. The consumers of IaaS are network engineers and administrators who enable an IT infrastructure, usually on behalf of the enterprise that employs them.

IaaS allows the most control over a cloud-based infrastructure. Much of the selection, configuration, and operation of the compute, storage, database, network, and other functions are controlled by the customer. However, the installation and maintenance of the underlying hardware is the responsibility of the cloud provider, allowing the consumer to focus on the operation of the infrastructure rather than the time-consuming and costly purchase, installation, and maintenance of their own hardware and software.

1.5.4 *"Anything-as-a-Service (XaaS)"*

Virtually anything can be offered "X-as-a-Service" with "X" being replaced by the letters representing the service. In other words, instead of the customer having to acquire and implement hardware and software, the application is implemented on that cloud provider's premises and accessed remotely by the customers via the Internet. Other XaaS examples are as follows:

- Containers-as-a-Service (CaaS)
- Storage-as-a-Service (STaaS)
- Functions-as-a-Service (FaaS)
- Database-as-a-Service (DbaaS)
- Network-as-a-Service (NaaS)
- Monitoring-as-a-Service (MaaS)
- IT-as-a-Service (ITaaS)
- Disaster Recover-as-a-Service (DRaaS)

1.6 Data Centers

Modern companies today cannot compete without extensive IT capabilities to process and store massive amounts of data. These capabilities are accomplished in data centers.

Data centers range from very small to huge facilities housing thousands of servers and related equipment (Fig. 1.2). Some data centers require an enclosed space of over one million square feet to house hundreds of thousands of servers. For example, the average Google data center houses over 900,000 servers (Galov, 2022). Connecting all those servers may require thousands of miles of fiber optic cables!

Building a data center is expensive and time-consuming. Real estate must be acquired. Hardware and software must be purchased, configured, and installed. Electricity and air conditioning systems must be provided. Skilled people have to be hired to install, operate, and maintain the data center.

Fig. 1.2 One of many Facebook data centers

1.7 Cloud Components

Regardless of size, there are four primary IT components in any data center: compute, storage, database, and networking.

1.7.1 Compute

In the past, computing was accomplished using mainframe computers. They are designed to service one or a few users that need to solve problems that require a lot of computation using a single or small number of applications. Solving large problems on mainframe computers may take hours or days to complete.

In a modern data center, the focus is on serving a large number of users from both the Internet and within the enterprise to quickly access numerous software applications that require, in comparison to mainframe computers, minimal computation. The computing platform to accomplish this is called a "server." It is much like a computer workstation but much more powerful. Whereas workstations may have a few to several CPUs, a server can have dozens to hundreds of CPUs as well as a very large amount of RAM storage. Servers are installed in equipment racks (Fig. 1.3). Besides computing, a server is often the platform for specialized functions such as a database.

Fig. 1.3 Data center servers mounted in racks

1.7.2 Storage

Data center operation requires a large amount of data storage capability. Just like your workstation or laptop computers, both temporary and long-term data storage is needed. Whereas RAM is quick and used for temporary storage, there are a variety of storage systems available for long-term data storage including hard disks, solid-state devices (SSDs), Small Computer System Interface (SCSI), tape drives, and caches. Depending on the needs of the applications, several different storage devices may be required.

1.7.3 Database

Server applications may need to access data, for instance, customer information, which is stored in databases. You can imagine a database as a high-powered spreadsheet. Again, depending on the needs of the application, one of several database types may be used in the data center. One or a combination of relational, non-relational, graph, document, and other database types may be required.

1.7.4 Networking

The data center must be able to communicate with their customers. If the customer has a website or other application running on the cloud provider's resources, it must also be accessible by *their* customers. This typically occurs across the Internet, but other more reliable networking options, such as private dedicated network connections, are available.

Internal networks that support the data center between physical and virtual devices can be very large, often supported by hundreds, even thousands, of miles of fiber optic cables. Since a single website request can generate requests for other devices within the data center, the bandwidth required within the data center often far exceeds that connected to the outside world.

1.8 Cloud Adoption Drivers

The benefits of the cloud over on-premise data centers are numerous. However, most of the focus has been on the possible economic benefits. The reality, contrary to popular perception, migrating to a public cloud provider may not result in cost savings. Therefore, cost savings should not be the primary selection criteria but the ability of the cloud to enable or improve business objectives. At minimum, categories that should be considered include business:

- Agility
- Adaptability
- Continuity
- Security
- Economics

1.8.1 Business Agility

Business agility is the ability to rapidly adapt to customer demands. A public cloud provider contributes to business agility in several ways.

Enabling traditional IT infrastructures often requires weeks, even months, of effort including equipment must be installed, cabled, configured, provisioned, and tested. On the other hand, public cloud providers have resources that are fully operational, on standby, which can be automatically and rapidly enabled. A global infrastructure can be turned up and fully operational in minutes.

A cloud infrastructure can adapt quickly to changes in demand for cloud resources through *autoscaling*. As the demand for service increase, referred to as "scaling up," autoscaling will enable more resources; as the demand decreases, the now-unused extra resources are disabled.

Another agile capability of a public cloud infrastructure is *load balancing*. Load balancing eliminates congested areas of a cloud infrastructure resulting in faster service. Load balancing can be implemented locally or globally.

1.8.2   Business Adaptability

In this case, business adaptability refers to the ability of the business to meet customer service expectations.

Major public cloud providers, because of the depth of their service portfolios, are positioned to meet almost any need that a customer might require. Services are offered far beyond the basic categories of compute, storage, database, and networking, but extend to hundreds of other applications such as Artificial Intelligence and Machine Learning.

Major cloud providers are adding additional services at a rapid rate to meet constantly evolving customer needs.

1.8.3   Business Continuity

Business continuity is the ability of the business to continue operating regardless of circumstances and should be of primary importance when considering migrating to a public cloud infrastructure. Three areas that are important are cloud redundancy, availability, and durability.

1.8.3.1   Redundancy

Cloud redundancy refers to how well cloud resources (both physical and virtual) are duplicated and to what degree customer data is backed up. Redundancy is usually initiated automatically after a failure. When a primary system fails, traffic is switched to the still-operating redundant system.

Redundancy in a public cloud provider's data center is implemented locally in a single data center as well as geographically where multiple distantly separated data centers are duplicated.

Redundancy is expensive. Very seldom can a customer afford to implement the levels of redundancy available from a public cloud provider where, in many cases, geographical redundancy is provided at no additional cost to the customer.

When redundant components keep the system operational when a failure occurs, the result is often referred to as *disaster recovery*. Disaster recovery encompasses surviving any event that threatens the availability of the public cloud provider.

1.8.3.2   Availability

The level of redundancy provided determines the availability of a public cloud provider. In other words, how sure am I that I can access the cloud provider from my location? Most public cloud providers provide system availability of 99.99%, or "four nines" of availability, which allows up to 52 minutes of downtime per year.

Actually, in the world of availability, four nines is not very good. However, because the public cloud provider has to consider that the network connection from the enterprise to the cloud provider (e.g., AT&T and Verizon) is not under the cloud provider's control, the availability has to be lowered considerably.

1.8.3.3   Durability

Durability refers to how available data is in the public cloud provider's infrastructure. Data is often highly replicated, and in one case, data duplicated six times across three geographically separate regions. Durability of data in most public cloud providers is 99.99999999% ("eleven nines"). This means that once across the connecting network to the cloud provider, the chances of your data not being there and accessible are very small. So small, in fact, that statistically, one failure of a few bytes of data will only occur once in thousands, if not millions, of years.

1.8.4   Business Security

Securing IT resources is more challenging than ever before. An advantage of migrating to a public cloud infrastructure is that much of the security responsibility shouldered by customers is now shared with the cloud provider, which has extensive security engineers, systems, and tools. Many security tools are accessible to the customer. Encryption of customer data is available at many levels of the cloud provider's infrastructure.

1.8.5   Business Economics

Migrating an enterprise IT to a public cloud provider can be financially beneficial. Some report savings of 50% or more. This is possible by trading capital expenditures such as hardware and other resources required on-premise for the reduced operational expenses of managing the cloud infrastructure. In addition, since resource capacity is flexible, customers pay only for the resources they use. Therefore, the cost of overprovisioning to meet occasional peak demands is avoided.

1.9   Virtualization Background

Modern cloud operation depends on **virtualization**. Through virtualization functions previously performed in hardware can be accomplished using software. Using virtualization, multiple virtualized computers, or **virtual machines (VMs),** can now run on a single hardware server. Each VM is referred to an "instance" of a virtualized computer (Fig. 1.4).

Computer virtualization is not a new idea. IBM used virtualization in mainframes as early as the 1960s. It was not until the evolution of the Internet and the transition from mainframe to server-based computing were accomplished that the need for its benefits was again felt. In a legacy data center example, each hardware server might only run one (or a few) web servers. Now, with virtualization, a single server could host dozens, even hundreds, of virtual servers. The result was a huge cost savings benefit for data centers. Instead of having to buy more and more servers, operators could perform more work using fewer costly hardware servers. Without virtualization, an economically feasible cloud would not be possible.

Virtualization is not limited to computing. It can be applied to web, application, database, and many other functions.

1.9.1   Hardware Server Operation

Before looking at how virtualization is accomplished, we need to look at the basic components of a hardware server (Fig. 1.5). Like any workstation or laptop, a hardware server requires processing capability (CPUs), an operating system, memory, and storage. Applications can then be loaded on the server to provide services to the server users.

1.9.2   Hypervisors

To accomplish server virtualization requires the use of a "hypervisor." It creates VMs and allocates *dedicated* hardware server resources such as CPU and memory to each. Each virtual server can be assigned more or less resources depending on

Fig. 1.4 (**a**) Single application server. (**b**) Virtualized server

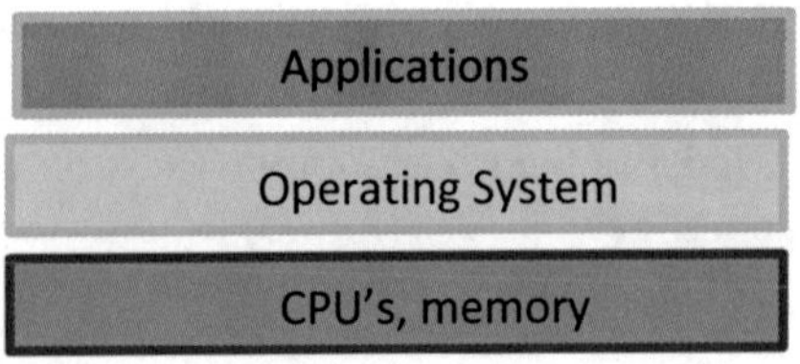

(a)   (b)

Fig. 1.5 Hardware server components

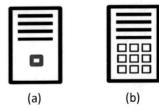

their need. This process allows each VM to operate independently of the hardware server within the confines of their allocation without overtaxing the physical server.

There are two types of hypervisors: Type 2 and Type 1.

1.9.2.1 Type 2 Hypervisors

Type 2 hypervisors sit on top of the hardware server operating system, such as Microsoft's Windows or Linux, and are called the *host* operating system (Fig. 1.6). When the hypervisor creates a server VM, it is given a scaled-down operating system called a *guest* operating system. The VM operating system can be different from the hardware server operating system. For instance, if the hardware server operating system is Windows, the VM can operate using Linux.

A disadvantage of a Type 2 hypervisor is that it still requires the use of a hardware server host operating system, which can be expensive, may cause delay, and has to be maintained. For these reasons, a Type 2 hypervisor is not practical for large operations but is adequate for personal computer use where one other operating system is desired to be used in addition to the host operating system. Popular Type 2 hypervisors include Oracle's VirtualBox and Microsoft's Virtual PC.

1.9.2.2 Type 1 Hypervisors

In contrast, Type 1 hypervisors do not use a hardware server operating system and therefore are referred to as "bare metal" servers (Fig. 1.6). Hardware servers that use Type 1 hypervisors can create dozens to hundreds of virtual machines using a menu of different operating systems. They are used extensively in a data center environment. Popular Type 2 hypervisors include Microsoft's Hyper-V, VMware's ESXi, and Linux KVM.

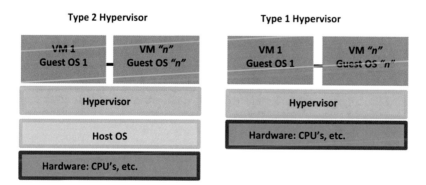

Fig. 1.6 Types 1 and 2 hypervisors

1.9.3   Virtual Machines

Each virtual machine (VM) created by the hypervisor has all the components needed to operate as an individual computer including an operating system, called a "guest" operating system. Each VM is allocated hardware resources such as CPUs and RAM by the hypervisor.

1.9.4   Virtualization Advantages

Virtualization resolves many of the problems that hindered the legacy data center design model.

1.9.4.1   CPU and Memory

Due to computing resource sharing and other conflicts, the legacy design model operates best with a one-to-one pairing between a hardware server and, for instance, a web server. Virtualization allows many VMs to be allocated their own slice of those resources without creating operational conflicts.

1.9.4.2   Operating System Dependency

Hardware servers in the legacy data center design model can run only one host operating system, for instance, Microsoft Windows. All applications installed must use Windows. If an application requires Linux, it must be installed onto another hardware server that runs Linux.

Virtualization using a Type 2 hypervisor actually eliminates the host operating system. VMs installed can independently run any guest operating system.

1.9.4.3   Scalability

"Scaling" means the ability to expand. To add capability to the legacy data center requires adding costly equipment that may take weeks to install and configure. Using a Type 1 hypervisor, VMs can be added or deleted automatically. Large data centers may create and delete thousands of VMs daily.

1.9.4.4   Redundancy

Providing redundancy in a legacy data center is expensive. It is not financially feasible to provide a backup hardware server for every other server. Therefore, choices have to be made about which legacy data center components are critical and require redundancy. Many servers will likely not have a backup redundant system due to cost. Even if a hardware server fails and a spare is available, it is a manual process to replace and make it operational. The result in any case is more down time. The only question is for how long.

In contrast, redundant software-based virtual machines can quickly and easily be created that are on different servers. The two servers can even be separated by large geographical distances. If a virtual machine fails, the other virtual machine continues serving with little or no down time.

1.9.4.5   Mobility

In a legacy data center, moving a web server from one hardware server to another may not be feasible. Therefore, if maintenance has to be performed on the hardware server, then that web server will be unavailable. However, VMs running on hardware servers, as software, can be moved to another hardware server quickly and easily without any disruption of service.

Another need for the mobility of server functions is operational efficiency. It may be advantageous to move similar VMs closer to servers that applications are running on or to balance the number of VMs running across multiple hardware servers.

1.9.4.6   Security

Security is one area where virtualization may not offer improvement but may be even more difficult. With the introduction of any new technology comes the possibility of greater security vulnerability. The features required for adequate security are usually provided by VMs where much of the security is the responsibility of the enterprise.

1.9.4.7   Economics

The economic benefits of virtualization are tremendous first due to far fewer hardware servers. A secondary benefit is reduced labor cost required for the predominantly labor-intensive data center management required to install and maintain legacy systems. A softer cost benefit is minimizing down time through increased redundancy, which results in better customer service preventing customers from leaving due to system unavailability or excessive delays.

1.10   Legacy Three-Tier Website Architecture

A common cloud data center use case is an e-commerce website. Customers access the website to search for and purchase a product.

Fig. 1.7 Three-tier
website (logical model)

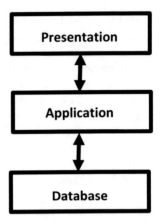

Traditionally, web application development has been based on three tiers (Fig. 1.7). The first is the "presentation" tier that refers to the user interface and includes HTML, CSS, and JavaScript. Second, the "application" tier, sometimes also called the "business logic" tier, is where data is processed. Finally, the "data" tier is where data is stored and managed. In general, platforms that relate to each of the above are the web, application, and database servers, respectively.

This three-tier approach is a logical model. In other words, it represents functions rather than physical devices. It is effective when developing software. However, it is not adapted to representing data center functions. Therefore, another model originally used by Amazon Web Services (AWS) is used instead.

In contrast to the logical model, this physical model defines four tiers: compute, networking, database, and storage. Understandably, this can be confusing since the reader may not be aware of the two different perspectives.

I will use the physical model for explanations in the rest of this book. These four components will be used to present the architecture of a data center using traditional methods. Others methods available build on the traditional model and will be discussed later.

1.10.1 Tier 1: Compute

Servers are the computers of the website. Just like your workstation or laptop computer, they can run many different software applications.

Two compute functions performed in a data center on hardware servers are a *web server* and *application server*.

1.10.1.1 Web Servers

To enable our e-commerce website, we will install and run web server software on a hardware server (Fig. 1.8).

Fig. 1.8 Web server

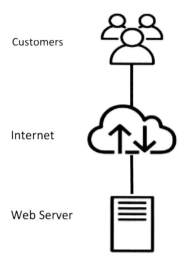

Customers

Internet

Web Server

Web servers are software applications installed that create web pages. Users access a web server using the HTTP protocol and view the returned information using a web browser such as Firefox or Chrome. The returned information might be a text, image, audio, video, or information from an application.

Traditionally, one software web server would be installed on one hardware server. If more traffic requires more web server capacity, more web servers were installed on a one-for-one basis on additional hardware servers.

Although a physical server is usually capable of running more than one virtual web server at a time, doing so can cause a lot of problems. For instance, if the physical server fails or requires maintenance, all of the virtual web servers will be inoperable.

Hardware servers can be obtained from many different companies including Dell and HP. Commonly used web server software products are Apache and Nginix.

1.10.1.2   Applications Servers

A web "application" is software that performs a specific function. For example, Microsoft's Outlook is a cloud-based email application. The number of software applications available is numerous and, like Outlook, is offered to customers for a fee.

Application software is installed on a hardware server (Fig. 1.9). Similarly, using traditional methods, only one (or a few) applications would be installed on a hardware server for the same reasons already discussed.

One problem is different applications require different operating systems. Some run on Windows, others on various versions of Linux. However, a physical server can only run one operating system at a time. If applications require different operating systems, separate hardware servers must be used.

Notice that both web and application servers are part the compute tier.

Fig. 1.9 Application and
web servers

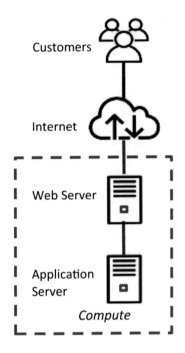

1.10.2   Tier 2: Networking

Networking, in particular the Internet, is a critical component of the web, data centers, and the cloud (Fig. 1.10). Not obvious is there are also extensive networking functions that operate within the hardware servers as well. Networking is discussed in much more detail later.

1.10.3   Tier 3: Databases and Storage

In the beginning days of the public Internet, websites were "static." In other words, the information they presented seldom changed. For instance, basic company information such as "About Us" was sufficient. However, modern e-commerce and other web applications require information presented to the user be updated in real time such as when a bill is paid online the account balance should be updated immediately.

1.10.3.1   Databases

Dynamic information is stored in databases (Fig. 2.5). Login credentials, user account preferences, purchase history, inventory, and wish lists are all examples of information that needs to be accessed by the application server for each website customer.

Commonly used database products include Oracle and MySQL.

Fig. 1.10 Data center networking

1.10.3.2   Storage

Just like a workstation or laptop, data center hardware servers need long-term storage capability (Fig. 1.11). Types of storage devices include traditional hard disk drives as well as newer solid-state device (SSD) storage.

Popular data center storage products used by public cloud providers include Dell EMC.

1.10.4   Legacy Three-Tier Website Example

Assume the above configuration is built by a small specialty bookseller. You enter the website name in your browser and access the bookseller's website via the Internet, which connects to the application server that contains all the information you would expect to see from a bookseller website. You enter the topic of the book you are interested in and the application server requests all titles that match from the database. The number of books of that title that are available, their prices, and other information are returned to the application server, presented to the web server, and then sent across the Internet using HTTP back to you.

Finding the book you want you click on "Buy now." When received, the database replies with a payment page that presents you with payment options where you enter your credit card number. The database processes the credit card and

Fig. 1.11 Cloud database
and storage

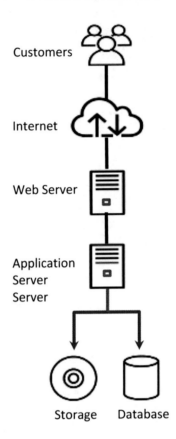

Customers

Internet

Web Server

Application
Server
Server

Storage   Database

returns the payment information to the application server, then web server, where it is transmitted back to you over the Internet with the message "Thank you for your purchase."

At this point you are done. However, your order is now forwarded to storage where the shipping department will retrieve the order information with the name of the book purchased and customer address. They will then locate the book, package it, and ship it to you. When completed, the database will reduce the inventory of the book by one and place a copy of your paid order in your "Orders" page.

All of these transactions occur at lightning speed across the Internet and the bookseller's internal network connecting the web server, application server, database, and storage system.

The example above is assuming a very small data center implementation. In reality, data centers are usually much larger. Obviously, our previous bookseller example is rather simple. Imagine the scale required to support Amazon.com. It has many data centers distributed globally, each with tens of thousands of servers in use.

1.10.5 Completed Architecture

Now assume the website business is growing. To handle the extra incoming Internet traffic, they have to add more hardware servers with web server and applications software installed on each. However, they discover that some orders are being delayed for some customers but some customer requests are getting timed out.

The problem is the traffic is accessing the first web server first. As the traffic load gets heavier, it starts to slow down. The other servers are not receiving customer order traffic. The solution is simple. The bookseller needs to add a hardware *network load balancer* in front of the hardware web servers (Fig. 1.12). The network load balancer will send equal traffic to each of the three hardware web servers.

Fig. 1.12 Application and network load balancing

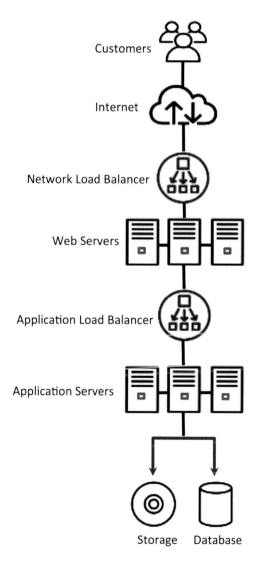

The extra traffic also occurs on the application servers. Similarly, a hardware *application load balancer* is installed in front of the application servers. Now customer traffic will be shared equally on all of the application servers.

1.10.6 Problems with the Legacy Model

This conventional, one-for-one relationship between hardware and the software servers that run on them had many disadvantages.

1.10.6.1 CPU and Memory

Although it was possible to run more than one web server application on a hardware server, doing so presents several problems. Although a server is a high performance, specialized computer, they have similar characteristics to the workstation or laptop we use every day.

Servers, like any other computer, have CPUs and RAM. Web and applications servers, databases, and other software are installed on the physical servers. They all require access to the computer's resources. This is a problem. Imagine a busy street with no stoplights. There will be delays and accidents. The same scenario applies to our physical server. Housekeeping chores and background processes in the physical server may take priority over serving Internet traffic. Delays will result in lost revenue when impatience drives customers to other websites to get what they want. To avoid this, only one (or a few) software server is often installed on a hardware server as to not overburden the server, to make management and maintenance easier, and to minimize the number of customers affect by failures.

However, this scenario is very inefficient. Hardware servers have the computing capability to run many software servers. Therefore, physical servers were drastically underutilized. Often, hardware servers that are not virtualized operate at 10% of capacity or less (Donovan, n.d.).

Hardware servers are expensive; operating like this has a huge cost. It is like buying a Ferrari that you can only use to travel at 20 mile per hour. Just as in this Ferrari example, you are spending a lot of money for performance you will not be able to use.

1.10.6.2 Operating System Dependency

Another problem with legacy data center design is hardware servers, just like your workstation or laptop, run on an operating system like Windows or Linux. While a hardware server can run only one operating system at a time, an application you want to use may require a different operating system. Therefore, servers also have to be dedicated to different operating systems resulting in even less operational efficiency and more cost.

1.10.6.3 Scalability

As a company grows, they must expand their data center capacity. Even a large hardware server has limitations on how many requests they can handle. To service more orders, additional physical hardware servers will need to be installed. This is a time-consuming process. The hardware must be ordered and received. Once acquired, they must be installed and cabled. Software must be purchased as well and loaded and the server must be configured for proper operation. Finally, they must be tested and brought online.

Responding to rapid increases in data center demands cannot be accommodated quickly. This is a process that routinely takes weeks or months. The company must predict far in advance what their demand will be to prepare far in advance for any expansion.

1.10.6.4 Designing to Peak Demand

Legacy data centers must design for the anticipated maximum demand. For instance, if your website sells a product whereby much of your annual revenue will occur in December, you must provide enough server and other capacity to fulfill those orders in a timely manner; otherwise, you will lose customers.

1.11 Virtualized Three-Tier Website

Let us see how virtualization can improve our bookseller's data center operation and decrease related costs.

The bookseller's business is successful and growing rapidly. They currently have hardware server each with web server software installed and running, which cannot keep up with the orders coming in. Adding more hardware servers is a problem. First, they are expensive. Second, they have to be ordered, installed, and cabled, software loaded, configured, and tested.

They quickly realized they need help. They hear of a smart IT person that is available and they hire her. She evaluates the situation and immediately recommends virtualizing the hardware servers. Type 1 hypervisor software is acquired and installed on each. She estimates that adding eight web and application servers will handle the current traffic with ease and so she adds eight virtual machine instances on the first web and application servers (see Fig. 1.13).

The IT person has just increased the number of web and application servers to 11 each. In addition, all incoming web traffic is evenly balanced between all the web server keeping their operation efficiency high. The bookseller has avoided spending thousands more on hardware servers and has immediately provided for the ability to rapidly receive and process orders improving their customer experience and preventing them from going to another bookseller to get their book.

Fig. 1.13 Website
architecture with
virtualization

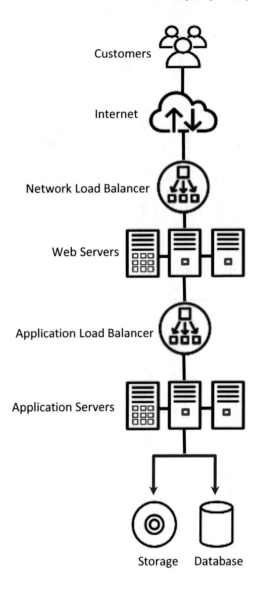

1.11.1   Redundancy

A company has to consider business continuity concerns. Failures will occur. Even short outages of a company's website will have a significant impact on customers and company revenue. How will the company respond to failures and continue to serve customers?

The solution is redundancy. Backup systems must be in place to takeover for failed equipment, which requires duplicated equipment, perhaps even a duplicate data center at a geographically different location is necessary.

Few companies can afford the cost of 100% redundancy. Therefore, they may only be able to recover from limited failures, which places their business at risk. In addition, replacing failed servers with backup servers often must be accomplished manually; hence, even with backup systems, the time to restore delay can be significant.

1.11.2   Mobility

Legacy data center installations are rigid. For instance, increasing the capacity of a server or moving applications to other hardware server to do maintenance on the former hardware server may be necessary. However, this is a manual and time-consuming process and is prone to failures. The result is less efficient data center operation.

1.11.3   Security

With major data breaches on major companies occurring daily, it is clear that the security of the data center is of prime importance. However, to properly secure a data center often requires many skilled and highly paid people. Even then cyber-criminals are exploiting as yet unknown system weaknesses. Many data centers are not equipped to properly secure their environment.

1.11.4   Cost

Common to all of the above disadvantages of a legacy data center is cost. Real estate, electricity and air conditioning facilities, hardware, software, and labor are just a few of the expense categories.

1.12   Summary

Cloud computing, as defined by NIST, includes five essential characteristics, three service models, and four deployment models. The decision to migrate from an on-premise infrastructure to a public cloud provider infrastructure should be based on how the cloud model meets business adaptability, agility, continuity, and security objectives and not just on cost savings.

Virtualization provides many advantages over traditional IT infrastructures. It is accomplished using a hypervisor, which creates virtual machines and allocates hardware resources they require.

A traditional, or "legacy," website is a common cloud application. I can reside on-premise or be hosted in a cloud data center. It was represented a physical model with compute, database, storage, and networking components. However, a legacy website application has numerous problems that are solved by virtualization discussed in the next section.

The hypervisor allocates dedicated hardware resources to each VM. There are two types of hypervisors: Type 2 and Type 1. Type 2 hypervisors still use a host operating system, such as Windows, and are used for smaller workloads. An example is VirtualBox on personal computers.

Type 2 hypervisors are not suited for data centers. Instead, Type 1 hypervisors are used to create virtual machines. A host operating system is not used on the hardware server; each VM is provided with a smaller guest operating system.

Homework Problems and Questions

1.1 Define and discuss the following.

(a) Five essential characteristics of cloud computing as defined by NIST.
(b) Three service models of cloud computing as defined by NIST.
(c) Four deployment models of cloud computing as defined by NIST.
(d) Four deployment models.

1.2 Define the primary cloud IT components of a data center.
1.3 Discuss the primary decision criteria that should be used to determine if moving to the cloud is beneficial.
1.4 Compare and contrast Type 1 and Type 2 hypervisors.
1.5 Discuss the advantages that result from using virtualization.

Bibliography

Donovan, P. (n.d.). *How virtualization can lead to hot spots – And what to do about it* [Online].
Fleera. (2022). *Flexera-State-of-the-Cloud-Report-2022*. Flexera.
Galov, N. (2022). *19 data center statistics you shouldn't miss in 2022* [Online].
Gartner. (2023). *Cloud infrastructure and platform services* [Online]. Available at: file:///C:/Users/01415036/Downloads/Markets_Cloud%20Infrastructure%20and%20Platform%20Services.pdf
Mell, P., & Grance, T. (2011). *The NIST definition of cloud computing*. National Institute of Standards and Technology.
Rappa, M. A. (2001). The utility business model and the future of computing services. *IBM Systems Journal, 43*(1), 1.

Part II
Cloud Networking

The modern cloud depends on networking. If it were not for fast, reliable networks, the cloud would not be possible.

Networking technology is broad, complex, and constantly evolving. To participate in the operation or engineering of cloud solutions, either public or private, requires a thorough understanding of basic TCP/IP networking technologies and the Internet.

Networking is present at many layers of the cloud. First, user connections to the cloud occur over a local area network (LAN) and then a wide area network (WAN), usually the Internet. The cloud provider's data center also has a complicated and extensive network. You may or may not be involved with the technical side of these networks. The last network is the cloud provider's data center network called the virtual private cloud (VPC). It is the network that connects all your virtual machines and cloud services together. It has all the properties of any other network. As a cloud engineer, it will be your responsibility to design, configure, and maintain it. Much of the cloud engineer's job is network engineering!

This section presents the minimum networking skills you will need to complete the labs and participate in the operation of the cloud!

- *Chapter 2: Basic Networking*
- *Chapter 3: Network Addressing and Protocols*
- *Chapter 4: Bringing It All Together*

The topics in this section are complex and can be difficult to learn. As you study this material, don't worry about learning it in detail right now. Get a good overall idea of the concepts, what they are for and how they are applied. Much more time will be required to master them if you decide to further pursue cloud skills or certifications. If you continue toward becoming a cloud engineer and, after completing the labs, know which cloud provider you want to focus on, you can come back and learn the material in detail.

So, remember this is difficult material. You don't need or want to be an expert right now. Take your time and don't get discouraged. It will make more sense after completing the hands-on labs. You can come back later for a closer look.

In the meantime, be like a tourist and enjoy the scenery!

Chapter 2
Basic Networking

2.1 Introduction

As a public cloud provider customer, you will interact with the cloud via a portal to the provider to create your own virtual private cloud (VPC), or space reserved for you in the cloud provider's infrastructure. This is where all of your compute, storage, database, and network resources will reside.

As the cloud engineer/administrator, it will be your responsibility to create and maintain the VPC network. Therefore, you must have basic networking knowledge. This chapter provides a summary of the basic industry knowledge and technical skill needed to pass entry-level cloud certification exams and to participate in an operational cloud environment.

Specifically, this chapter provides you with an overview of the networking industry history. This will help you understand some of "why" things are done the way they are in the cloud. Next, an overview of the OSI seven-layer model will show "how" the network is connected and operates.

Much of the material that follows is detailed and may be boring. However, it is vital that if you want to participate in the cloud world that you persevere and learn.

2.2 The "Network"

For many years, the "network" could only connect telephone calls. Today, it is hard to comprehend a world without personal computers and networks capable of carrying billions of bits of data globally in microseconds. Although commonplace today, the evolution from voice to data networks has been tedious.

© The Author(s), under exclusive license to Springer Nature Switzerland AG 2024
M. S. Kingsley, *Cloud Technologies and Services*, Textbooks in
Telecommunication Engineering, https://doi.org/10.1007/978-3-031-33669-0_2

2.2.1 Telephone (Voice) Networks

When Alexander Graham Bell's patent expired for the telephone, it opened the door for entrepreneurs to build networks and offer telephone services (National Archives, 2020). Each new provider built their own network with their own telephone wires, which were strung on poles high above the ground. In some places, there were so many wires the sunlight did not reach the sidewalk (Fig. 2.1).

Regardless of which telephone network a customer chose, they could not communicate with everyone, only to those connected to that network. As a result, Theodore Vail, the president of American Telephone and Telegraph (AT&T), saw an opportunity. He proposed that the US federal government that the telephone industry should be a monopoly. If accomplished, all customers would be connected to one network that could provide "universal service" to all customers. In other words, everyone would be able to seamlessly communicate with everyone else across the country. It has been said that "Alexander Graham Bell invented the telephone and Theodore Vail invented the telephone business" (Turner, n.d.).

An agreement was reached and AT&T (with a few exceptions such as in rural areas) was granted a monopoly over telephone industry in the United States. In return for their investment, AT&T was guaranteed as lucrative profit.

AT&T built an extraordinary national network. It is one of the greatest engineering achievements of the twentieth century. However, calls on the network were analog. Analog transmission suffered from many problems. Over distance calls lost their strength and had to be amplified. The further the call had to travel, the more the

Fig. 2.1 Bell Telephone
System Wires circa 1900.
(The Evolution of
Telephone Cable, n.d.)

analog signal degraded. Long-distance calls were not only expensive but also unreliable and distorted.

It was discovered that digital signals were not subject to the problems of their analog counterparts. Since digital signals were "ones" and "zeros," or voltage that turned on and off, they could be regenerated easily. In contrast to amplification, regeneration could detect the digital signal, recreate a clean copy, and forward it eliminating any distortions. Long-distance calls were now crisp and clear.

AT&T was hesitant to evolve into digital technology. They had an analog network that was protected from competition that was very profitable. Adapting the network to digital would require millions of dollars of investment. Even though they were guaranteed a profit, they still balked. This initiated battle between AT&T and the US federal government demanding they upgrade their network digital and other technologies. AT&T did not realize that this stance and other behaviors were the beginning of the end of AT&T as a monopoly. An era of competitive forces against the company began.

2.2.1.1   Carterphone

AT&T relentlessly protected their monopoly stranglehold on the telephone industry. However, entrepreneurs were attracted by the challenge and the profit potential of telephone services and equipment. One such maverick was Thomas Carter, who owned a trucking company. He developed a device called the "Carterphone" (Carterfone, ATT and the FCC 1948–1967, n.d.) (Fig. 2.2). His idea was that if he connected one end of the tube to a telephone and the other to a ham or two-way radio, he could talk to a truck driver over a telephone connection. He also started selling his device to others.

AT&T did not allow not only the Carterphone to connect to their network but also any other device to be connected, using the defense that foreign devices would insert voltage into their network and cause damage. Although this was true, the Carterphone was not an electrical device.

Instead of allowing the Carterphone to be used, AT&T aggressively pursued Carter in court. What they did not realize was that Carter was a belligerent Texan and a man of principle. He refused to be intimidated by a large corporation. He spent his entire fortune fighting AT&T in court. Eventually he won the right to attach his device to the AT&T network, but he never recouped his financial losses. What Carter's battle did accomplish, though, was that it opened the floodgates of competition against AT&T.

Fig. 2.2 Carterphone

Telephone   Radio
Network   Network

2.2.1.2 Competition

AT&T viciously defended any competitive attempts on their domain. For instance, if you wanted a telephone, it had to be supplied by AT&T. You could not buy it, but could only rent it. There was only one model which used rotary dialing, and it was black (Fig. 2.3). Again, the objection by AT&T was that connecting a random device might inject voltages into the network and cause damage. AT&T was eventually forced by the court to allow non-AT&T devices to be connected to the AT&T network as long as they were tested and shown not to cause damage to the network. This applied not only to telephones but also any equipment that could be connected to the network.

Competition also came from companies trying to enter the telephone network business. By connecting cities with microwave radio towers that carried telephone calls, Microwave Communications, Inc. (MCI) was able to bypass the AT&T network. AT&T interpreted this as a direct challenge to their monopoly on telephone service.

Other companies such as Sprint also challenged AT&T. Vicious court battles between these and other ambitious competitors continued for years further motivating the United States Department of Justice (DOJ), which is responsible for preventing anti-competitive business behavior, relentlessly attempted to not only eliminate AT&T's monopoly but also dismantle the company.

Finally, in 1984, AT&T agreed to a settlement with the Department of Justice (DOJ). AT&T's long-distance business was lucrative, but the local service business was much less so. Based on thousands of miles of wire and connections to residences, the local network was maintenance intensive and much less rewarding. Therefore, AT&T proposed it would keep the long-distance telephone business and divest itself of the less profitable local telephone network. They agreed to break the local network service business into seven separate regional companies, which would provide local telephone service. Soon, long-distance competition from MCI, Sprint, and others would be allowed.

Eventually, AT&T did evolve to digital transmission. In the following years, more changes followed. In the past, cable TV companies were not allowed to offer

Fig. 2.3 Model 500 telephone. (500-Black, n.d.)

telephone service and telephone companies could not provide television service. In 1996, these regulatory barriers were also eliminated. Telephone companies as well as cable TV providers could offer the "triple play" of telephone, television, and Internet service.

Finally, since in the opinion of the judicial system the goal of competition in telephone service was achieved, AT&T was allowed to purchase back the seven divested regional companies. Today, competition in the telecommunications industry continues to be intense. New companies rapidly formed while others merged. Recently, Sprint, the last of the early pioneers that challenged AT&T, merged with T-Mobile.

2.2.2 Data Networks

In the voice-only network era, data on a network was an afterthought. There were no personal computers, only large mainframe computers. Connecting remotely to a mainframe computer could only be accomplished by dedicated, point-to-point network connections to a remote user by using a **modem**, which converted digital computer data to analog for transmission where the receiving modem would reverse the process to convert the analog signal back to digital that could be read on a monitor.

Since analog networks were prone to distortion and other problems, transmission lines had to be carefully engineered to carry data. They were very slow and expensive. The cost of creating a large mesh of connections to many endpoints was prohibitively expensive.

2.2.2.1 ARPANET

In the background, there was an effort headed by the Advanced Research Project Agency (ARPA) to build a packet-based network called ARPANET. Packet networks break data into discrete chunks for transmission. ARPANET was one of the first to refer to their network as a "cloud" because as packets entered the network, they were routed to their destination based on network metrics and conditions. Sequential data packets did not have to take the same route together to their destination. Therefore, the location of any specific packet could be anywhere in the network.

ARPANET was also designed to survive massive outages. If there were any network failures, the packets would be resent to their destination using an alternate path, therefore surviving the outage. In other words, the network was "self-healing."

Later, ARPANET would become publicly accessible. We now know it as the "Internet."

2.2.3 Telephone Versus Data Networks

Although the Internet was publicly accessible, it was in its infancy. It had evolved to include connections to hundreds of other networks in other countries across the globe, making end-to-end reliability impossible. It was designated on a "best effort" basis. If a packet was not delivered in a certain amount of time, it might or might not be resent due to a problem in one of the connecting networks. Also, being a public network, the Internet was inherently not secure. On the other hand, dedicated network connections where two endpoints were physically connected with wires were expensive, but reliable and reasonably secure. In the case of AT&T, there was a huge reluctance to offer packet-based services because it required building a completely separate packet network in parallel to the telephone network. Cannibalizing the lucrative dedicated network for a huge investment in packet network facilities just did not make business sense.

AT&T's strategy of maintaining current profit over offering more and better services to customers continued. However, as computing technology evolved, companies wanted the ability to move data between locations and competing network providers were forced to accommodate.

Customer demand motivated AT&T to build packet networks and offer packet services. However, the TCP/IP protocol suite used in the Internet had not been adopted by conventional network operators.

Protocols like X.25 and Frame Relay were used first. However, each data network improvement required another separate and distinct network to be built. There was a telephone network and numerous data networks. The "network" was actually many distinct and physically separate networks.

Later, the federal government mandated the use of Internet protocols be used in all federal networks, which resulted in virtually all networks, both public and private, being based on the TCP/IP.

2.2.4 "All-In" on Packet

Today, all telephone calls are converted to packets by using Voice over IP (VoIP). All major networks carry voice calls as VoIP data packets. The legacy voice telephone network is disappearing. The only remnant of the original analog telephone network is the "last mile," or wired telephone connections to residences. Of course, as more people are using cell phones rather than wired telephone service, this part of the network will decline even more. As a result, all voice, data, and video traffic carried on networks is packet-based and uses TCP/IP.

2.3   OSI Model

Any discussion on networking begins with the Open Systems Interconnection (OSI) model. Published in 1984 by the International Standards Organization (ISO), it is a conceptual framework describing how network systems communicate. The model is divided into seven layers (Open Systems Interconnection-Model and Notation, 1994) (Fig. 2.4).

A "host" is a generic term for any device at the endpoints of a communications link. Host devices include computers of all types (workstations, laptops, iPads, etc.), servers, and other devices. Only host devices operate using all seven of the OSI layers.

In contrast to hosts, even though the data from all seven layers is present, network devices only see at most the bottom three OSI layers. Network devices include switches and routers; switches operate at layers 1 and 2; routers at layers 1, 2, and 3.

Let us begin by looking at the general responsibilities of each of the OSI's seven layers for now. Details of each layer will be examined later.

2.3.1   Layer 7: Application

When the term "application" is used we often think of a program. For instance, Microsoft Word or Excel may be referred to as applications. That is not the definition of Layer 7, which is where network applications reside.

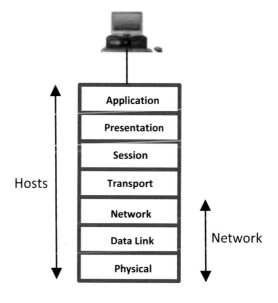

Fig. 2.4 OSI seven-layer model with host and network areas defined

2.3.2   Layer 6: Presentation

This layer is where data is translated and formatted. Characters are encoded and data compressed if required.

2.3.3   Layer 5: Session

Layer 5 acts like a moderator in that it controls the dialogue between network devices. It sets up and terminates communication sessions.

2.3.4   Layer 4: Transport

Layer 4 coordinates data transfers between hosts including how much data to send and the rate of data transmission. It also shares the communications channel, through a process called multiplexing, where different user applications take turns sending data. On the receiving side, in-order sequencing of packets is determined, and requests for retransmitting of data packets received in error are accomplished.

2.3.5   Layer 3: Network

Layer 3 is primarily responsible for host-to-host, end-to-end packet addressing and routing.

2.3.6   Layer 2: Data Link

The data link layer connects to physical network media via network interface cards (NICs). Each NIC provides local addressing that enables hop-by-hop connections between network devices.

2.3.7   Layer 1: Physical

Layer 1 has a lot of responsibility. It must interface with numerous media types and encode the bits so they can be transmitted effectively. For instance, to transmit data across fiber optics, the bits must be converted to light; for wireless, radio frequencies. Similarly, other media types require other encoding mechanisms.

2.3.8   Encapsulation

Each layer's actions are communicated using "encapsulation" (Fig. 2.5). Data enters the top of the OSI model at Layer 7, which performs its functions and puts instructions on what it did in a "header" that is attached to the raw data. Each layer in turn performs its functions and describes what has been done in that layer's header. This information added at each layer is generically referred to as "overhead." Finally, reaching Layer 1, the data stream is encoded for the type of media and then transmitted as binary bits. When the bit stream is received, the process is reversed. Each header with instructions is read by the receiver, and functions are accomplished. Finally, the original transmitted data is recovered by the receiving host user.

The diagram shows only the operation of the hosts, or end devices. The operation of the connecting network will be discussed later.

2.3.9   Comparison

An analogy can be made to the children's toy called "Russian dolls" (Fig. 2.6). The dolls are sequentially encased until all the dolls are contained in the largest doll's casing. Each doll and its casing can be thought of as an OSI layer with header information. The instructions are read and performed on the data from each casing.

2.3.10   Advantages of the OSI Model

The acceptance of the OSI model overcame many limitations in legacy networking.

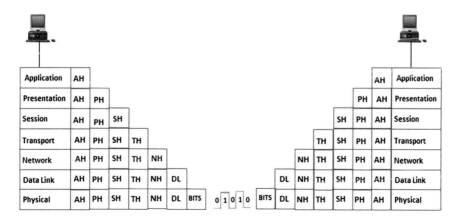

Fig. 2.5 OSI Encapsulation

Fig. 2.6 Barrel of monkeys

2.3.10.1   Vendor Interoperability

Prior to OSI model publication, there were very few standardized rules for how networks worked. Therefore, equipment manufacturers built products that were proprietary.

To communicate with a device from a particular vendor's equipment on the one end required using the same vendor's equipment on the other end. The result is referred to as "vendor lock-in" because the network provider was then at the mercy of the equipment manufacturer and had little leverage on product pricing or future feature planning. It was also a financial risk if a particular company were sold to another company, possibly even to a competing network provider, or declared bankruptcy.

2.3.10.2   Competition

Because of the OSI model, competition between network equipment vendors has shifted from proprietary functionality to open and standardized network protocols. Since different equipment vendors now interoperate, network providers are not held hostage by equipment vendors. They can mix and match different equipment in the network.

It is important to make clear that standards define how network equipment interfaces and operates in the network. In other words, standards dictate *what* the equipment must do, not *how* it does it. What happens inside the boundaries of the equipment vendor's product is up to them, proprietary, and (hopefully) to their competitive advantage. As long as the product "ins and outs," or network protocols used to communicate between the devices, are standardized, everything works fine.

2.3.10.3 Modular Engineering

Prior to the OSI model, software that instructed network equipment hardware on how to operate was developed without much structure. Granted, in the beginning, network software was relatively simple. As networks grew in complexity, so did the software supporting them. "Bugs," or mistakes, in software development are inevitable as is the need for improving and adding to the capabilities of network software. Making corrections and additions when software is built sequentially requires opening up all the software, making the changes, and then rebuilding it. This process can be time-consuming as well as may introduce other problems into the system.

The OSI model enables modular software development. Each layer can be thought of as a software module. Much like a dresser with drawers, each drawer has software that performs the functions of that layer. It is not necessary to open the software from other layers to correct problems or modify them.

2.3.11   Problem: Focused Skillsets

A disadvantage that a student learning cloud technology has versus an established network engineer is the level of technical networking knowledge and skill required. Network engineers are responsible for installing, configuring, and maintaining network equipment and connections. Unlike Microsoft, there are few Windows-like applications. Most connections to network equipment are via a command-line interface (CLI), and each vendor of networking equipment has their own proprietary CLI. If you have ever used MS-DOS and recall the "C:\" prompt you are familiar with CLI.

The advantage of using a CLI is that the user can complete tasks quickly. If you have ever had to drill down through six or more levels of folders in Windows to find a file you can relate. With a CLI, you could enter a command and go directly to the file. Although there is a trend toward a Windows-like interface for network equipment, all tasks including configuration and troubleshooting are generally still accomplished using the vendor's CLI. Most experienced network engineers prefer using a CLI to a Windows-type interface due to the speed and convenience they offer. However, this level of skill requires an expert level knowledge of network protocols and their functions.

Fortunately, this level of network expertise and skill is not necessary when creating and operating a cloud infrastructure with a public cloud provider. Since the cloud provider has "abstracted," or hidden, the network hardware accessing network functions does not require skill in using a variety of CLIs from different equipment vendors. Instead, the cloud provider offers a user-friendly portal interface similar to

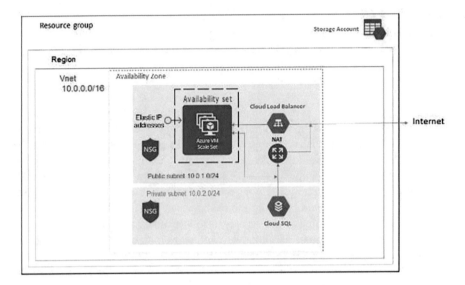

Fig. 2.7 Example Graphical User Interface (GUI)

a Windows environment (Fig. 2.7). Therefore, an expert-level knowledge of networking is not required of the cloud engineer. However, a thorough understanding of networking protocols, or communication rules and syntax, is necessary to interface with the network engineers to integrate the public cloud provider infrastructure with the customer network.

2.4 Network Protocols

So far, you have learned about the OSI seven layers and their functions. This chapter will extend that discussion to the network protocols that you will need to understand that operate at each of these layers from top to bottom.

Layers and their related protocols bolded in Fig. 2.8 are those of most concern to network and cloud engineers that will be discussed here in detail. However, for continuity, the protocols that are not bold will also be briefly introduced.

2.4.1 Layer 7: Application Protocols

Data, such as something you type on your keyboard, enters the top of the OSI "stack" at Layer 7, or Application layer. The Application layer is not like an "app" on your cell phone or a program like Microsoft Word. The protocols that reside here are *network* applications. Three protocols are most important to the cloud engineer:

Application	7	HTTP, DNS, DHCP
Presentation	6	JPEG, GIF, MIME, ETC.
Session	5	SOCKETS
Transport	4	TCP, UDP
Network	3	IP
Data Link	2	Ethernet
Physical	1	Media specific encoding

Fig. 2.8 OSI layer-related protocols

Dynamic Host Configuration Protocol (DHCP), Domain Name Service (DNS), and the Hypertext Transfer Protocol (HTTP).

2.4.1.1 Dynamic Host Configuration Protocol (DHCP)

Imagine you are plugging your laptop (a "host" device) into an Ethernet network connection. The first thing that needs to happen is your laptop needs an IP address. To get an IP address, it needs the help of a Layer 7 protocol called the Dynamic Host Configuration Protocol (DHCP). The Dynamic Host Configuration Protocol (DHCP) is a Layer 7 protocol that automatically assigns IP addresses to network devices. This process is working behind the scenes, you as the user are unaware it is happening. Once the laptop has an IP address, it can connect to the network.

2.4.1.2 Domain Name System (DNS)

You want to reach the website www.xyz.com. However, the network does not understand human languages (at least not yet). It needs the destination IP address of the xyz web server. Now you need the help of the Domain Name System (DNS).

When www.xyz.com website name is entered into an Internet browser, the computer has no idea how to get there. Instead, the computer accesses the DNS that resides in a remote server somewhere, and asks it what IP address www.xyz.com is at. DNS returns the destination IP address of www.xyz.com, which is then entered with the user's source IP address in the IP packet. The laptop user is unaware this process is occurring.

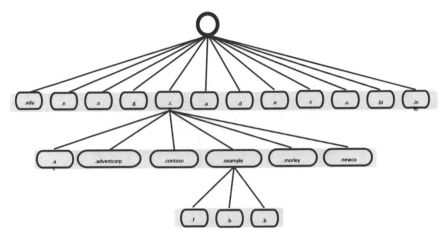

Fig. 2.9 Domain Name System (DNS) hierarchy (DNS Tutorial part 1 – DNS basics, n.d.)

The DNS system has a global hierarchy (Fig. 2.9). Most organizations will have a local DNS. Your computer queries it first. If the local DNS does not know the IP address of www.xyz.com, it will then ask a regionally located DNS in the hierarchy. If it does not know, it follows the tree upward. The last and final DNS server is the root DNS. If the root DNS does no know where to send the request next and other DNS cannot locate the IP address, then the webpage does not exist or it cannot be reached.

Now that you have an IP address of your computer (the source host) and the IP address of the web server (the destination host) you are trying to reach, you can now start to connect to the network by moving down the seven OSI layers to Layer 1. It begins with the Hypertext Transfer Protocol (HTTP).

2.4.1.3   Hypertext Transfer Protocol (HTTP)

An important Application layer protocol is the Hypertext Transfer Protocol (HTTP), which is a standardized protocol used to communicate with webpages. The encapsulation process begins with HTTP. It takes your data and formats it for transmission across the network. It then attaches overhead to the data in the Application Header (AH) that says "I have formatted this information using HTTP" and hands it to next lower layer, the Presentation layer.

2.4.2   Layer 6: Presentation Protocols

When receiving the data from the Application layer, the Presentation layer performs more formatting and other functions. For instance, a .jpg (and other formats such as .gif) picture, it will put information in the Presentation Header (PH) that says, "Hey, this is a .jpg picture." Similarly, it will format emails a Multipurpose Internet Mail Extensions (MIME). Other applications require other formats implemented in Layer 6. All data at this point is encapsulated by the Presentation Header.

In-depth knowledge of Layer 6 protocols is not usually necessary when working in the cloud even for network engineers.

2.4.3   Layer 5: Session Protocols

The Session layer's function is fairly straightforward and specific—it is responsible for opening and closing network session between hosts using a mechanism called "sockets." To work with sockets requires extensive programming skills. Again, the instructions on what the Session layer does to the data stream are encoded in the Session header (SH).

Much like Layer 6, detailed knowledge of Layer 5 is seldom necessary for network or cloud engineers when working in the cloud.

2.4.4   Layer 4: Transport Protocols

Working knowledge of the Transport layer is mandatory in the cloud environment. It has many responsibilities. First, it sets up host-to-host communications. It also guarantees reliable transmission of data; if data is lost in transmission, it initiates and completes retransmission of the missing data. It accomplishes "multiplexing," or the regulation of interleaving packet from different applications for transmission. For example, it may receive data from email, a web browser, or other applications. Each application's data is identified by a port number. For example, web browser data uses port 80 and email port 25. There are many port numbers that can be used depending on the application.

The Transport layer also takes care of making each application's data is evenly interleaved with the others so no application consumes all the available bandwidth. In other words, TCP makes packets take turns to be transmitted.

As with other layers, the instructions detailing what the Transport layer has done to the data stream is identified in the Transport Header (TH).

There are two important protocols used in Layer 4: TCP and UDP.

2.4.4.1   Transport Control Protocol (TCP)

The Transport layer has two modes of operation. The first is "connection-oriented." In this mode, a reliable and dedicated connection is established between hosts before sending packets. This is accomplished using the Transmission Control Protocol (TCP).

Another important function of TCP is "flow control." If the transmission media becomes congested with too many packets, the Transport layer will decrease the volume of packets being sent. However, real-time applications such as voice and video are very sensitive to delay caused by retransmission of lost packets. Pauses in a voice call or video are not acceptable to users. Therefore, another Layer 4 protocol is used when delay cannot be tolerated.

2.4.4.2   User Datagram Protocol (UDP)

The User Datagram Protocol (UDP) is a reduced functionality version of TCP. It is used for real-time applications such as voice calls or video streams. It uses the "connectionless" mode of operation, which does not set up a reliable connection between hosts before packets are sent because voice calls and video streams are relatively immune to the effects of a limited number of lost or out-of-order packets. The human ear and eye cannot detect such small changes in the packet stream. Therefore, UDP does not initiate packet retransmissions when packets are lost, does not reorder the sequence of out-of-order packets, and does not enforce flow control.

2.4.5   Layer 3: Network Protocols

More than any other layer, detailed knowledge of the Network layer is required to participate in the technical cloud. It is the most recognizable because this is where the Internet Protocol (IP) resides.

The Network layer's main responsibilities are addressing and routing. This information is carried in the Network Header (NH).

2.4.5.1   IP Addressing

Most people are familiar with IP addresses. They are used to identify network device interfaces. The network operates on IP addresses that are 32 binary bits long but represented using the dotted decimal formatting, which makes them easier for humans to read. An example is as follows:

192.10.10.1

IP addressing will be discussed in much more detail later.

2.4.5.2 IP Routing

The question is how do those endpoints get connected? The answer is routing, or determining what links in the network the data, now referred to as a "packet," needs to cross to reach its destination. However, first we need to introduce the network *router*.

2.4.5.3 Routers

Years ago, there was no Wide Area Networks (WANs) as we know them today. Hosts were connected using Ethernet in a Local Area Network (LAN). If data was to be transmitted to remote locations, it could only be accomplished by with dedicated connections using modems across the existing voice telephone network. Transmission was slow and error prone.

Telephone network providers developed and implemented early generation packet networks using various now obsolete protocols. Meanwhile, the US federal government was building the Arpanet, which later became the Internet, using TCP/IP. Therefore, to connect hosts on the LAN to others across a WAN required a router.

Routers are Layer 3 devices (Fig. 2.10). Even though packets they process include headers from all seven OSI layers, a router can only see and process the Layer 3 Network Header (NH) and Layer 2 Data Link Header (DH).

Information in the NH includes the source and destination IP addresses of the hosts on the originating and terminating end of the network that are trying to communicate. The source IP address is the originating host and the destination IP address is the terminating host. The router routes the packets through the intermediate network. The source and destination IP addresses do not change during their end-to-end trip.

The Internet Protocol has two versions currently in use. IP version 4 (IPv4) is older and was intended to be replaced by IP version 6 (IPv6). However, the transition has been slow. The infrastructure of most public cloud providers is slowly being retrofitted to use IPv6 but IPv4 is still often preferred.

IP routing can be accomplished in two ways: statically and dynamically.

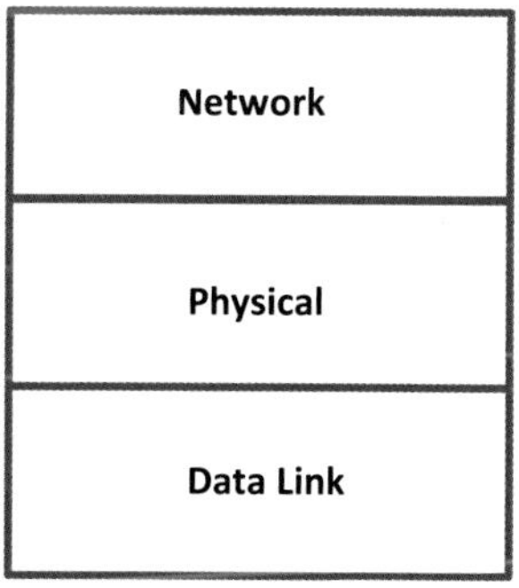

Fig. 2.10 Router OSI layers

| Network |
| Physical |
| Data Link |

2.4.5.4 Static Routing

Network operators can manually configure where packets are routed in a network. This is called "static routing." Using static routes is satisfactory for small networks. However, as the size of a network grows, manually creating static routes to allow a full mesh connectivity between all network routers increases their number exponentially.

To determine how many links are needed to create a full mesh network that allows all users to talk to each other using the formula $n(n-1)/2$, where "n" is the number of routers. With four routers, six links are necessary for a full mesh connection between all routers. However, if 12 routers are in the network, then 66 static routes have to manually created; for 20 routers 190 static routes; and 100 routers have 4950 routes.

As you can imagine, creating a lot of static routes will be time-consuming, error prone, and practical only for small networks. Thankfully, there is a way to route packets automatically.

2.4.5.5 Dynamic Routing

Packets can be routed automatically using dynamic routing protocols. Working in the cloud requires an understanding of how dynamic routing works and the protocols that can accomplish it.

The way dynamic routing protocols work is each network device (hardware or virtual machine) floods the network connections that it has to every other device. Each device then creates a map of the network.

Each network device has software that is used to calculate a route from every network device to every other network device. Before packets are transmitted, each device uses an algorithm to compute how to get to a packet's destination. This information is stored in each device in a "route table." Popular algorithms are least number of hops or the shortest path from the source to the destination device. Each device makes the decision of where to send the packet once it is received by referencing the onboard route table.

2.4.5.6 Dynamic Routing Protocols

There are several popular dynamic routing protocols used in the cloud. They include Routing Information Protocol (RIP), Open Shortest Path First (OSPF), and Intermediate System to Intermediate System (IS-IS).

Detailed knowledge of each dynamic routing protocol is not necessary in a public cloud provider environment. However, creating and maintaining your cloud infrastructure will require that you be able to configure routing tables in your virtual private cloud.

2.4.6 *Layer 2: Data Link Protocols*

While the Network layer is responsible for connecting hosts end-to-end across the network, the Data Link layer does so from network device to network device. While the Network layer IP address determines which links the packet will transverse from source to destination and never changes, the Data Link layer locally connects the network interface card (NIC) of neighboring devices. Each NIC has a unique Data Link address.

The Data Link layer functionality resembles a relay race (Fig. 2.11). Each device hands off the baton to the next sequential device.

Exactly like other layers, the Data Link layer encodes information and instruction on what it did to the data stream in the Data Link Header (DH). The package is now referred to as a "frame."

The Data Link layer is also responsible for error detection. If a receiving device detects an error in a frame transmitted it simply drops it. The Data Link layer makes no effort to recover the lost frame. If that is needed, TCP in Layer 4 (if used) will detect the missing frame and initiate recovery of the data.

In the evolution of networking, there have been many Layer 2 protocols used from generation to generation. However, Ethernet has now become the primary Layer 2 protocol used today not only for LANs but also, with modifications, in WANs.

2.4.6.1 Ethernet

One of the oldest network protocols still in use today is called Ethernet. It was developed by Robert Metcalf at Xerox in 1973 (Metcalfe, 2013). It was a Local Area Network (LAN) protocol, which connected devices, as the name implies, "locally," or in small area such as an office.

Fig. 2.11 Layer 2 handoff

The primary purpose of an Ethernet LAN is to connect all the locally installed host devices. Host devices are anything that can connect to the network. Computers of all type including workstations, laptops, iPads, and servers as well as printers, even gaming platforms are host devices that connect to the LAN. They can communicate only with other devices on the LAN.

When Ethernet was invented, there was no Internet or other public data network as we know them today. Connection between remote locations was via dedicated, not shared, connections using analog modems. It may be hard to fathom by some today, but there were also no personal computers! Mainframe computers were the primary computing platforms.

Connecting to a mainframe computer to a monitor (not a computing device, just basically a TV screen) was originally accomplished with local dedicated connections from monitors to a limited number of network ports on the mainframe. However, with Ethernet, many monitors could be networked together and connected to one mainframe network port. Since the mainframe ran much faster than the monitors and the human brains using them, it appeared as though everyone had a direct connection to the mainframe computer. In reality, the mainframe was "time sharing" its processor with everyone. It would spend a short amount of time working on each job.

The utility of Ethernet became obvious and its popularity grew rapidly. The responsibility for the advancement of Ethernet was taken on by the Institute of Electrical and Electronic Engineers (IEEE). Since then, there have been dozens of additions and modifications to Ethernet including adapting its use to the Wide Area Network (WAN). "Carrier" Ethernet has now replaced many previous Layer 2 WAN protocols previously in use. Efforts to further the advancement of carrier Ethernet is now accomplished by the Metro Ethernet Forum (MEF) (About MEF, 2023).

Ethernet Addressing

Ethernet devices found each other by using "physical" addresses. Each device, such as a workstation, has a network interface card (NIC), which connects it to the LAN. Each NIC is assigned a globally unique and unchangeable hardware address by the device manufacturer. The address is sometimes referred to as being "burned in."

Layer 2 addresses were referred to as "Media Access Control (MAC)" addresses. Each device NIC has a source MAC address. To reach another Ethernet device, the originating device would create a frame with its source MAC address and the MAC address of the destination, which is the NIC of the next switch or router. It then forwards the frame to the next network device NIC in the network.

MAC addresses are quite different from IPv4 addresses. First, they are 48 binary bits long. Just like IP addresses, people cannot readily read binary bits. To simplify being able for people to interpret a MAC address, Ethernet used 12 hexadecimal numbered groups, each letter or digit representing four bits, separated by dashes. An example is as follows:

$$68 - EC - F5 - 2B - D7 - A1$$

MAC addresses and hexadecimal numbering will be investigated in much more detail later.

Ethernet Broadcasts

Ethernet originally worked on the concept of "broadcasting" over a single cable connecting all of the local host devices. All of the Ethernet host devices on the network could communicate with all of the other host devices. When an Ethernet device wanted to send data to another Ethernet device, it would listen until there was no traffic on the shared network cable. It would then send a frame with its source MAC address and the destination MAC address onto the network. All other Ethernet devices on the network could see the frame. Only the Ethernet device that had the destination MAC address resident in its NIC card would accept the frame; all other devices would simply ignore it. After looking at the data, if a response was needed, the destination device would return a response frame with its source MAC address and the original sender's source MAC address as the destination MAC address.

Ethernet Hubs and Switches

Using Ethernet broadcasts had some problems. The obvious one was collisions. Just as in vehicle traffic, the more cars are trying to access a highway the more collision that will occur. Eventually, as more hosts tried to access the Ethernet network, the result was that no one would be able to access it because of all the collisions.

Another analogy describing Ethernet broadcast operation is when a politician holds a press conference. All the reporters scream and try to get the politician's attention to ask their question. Eventually the politician picks one reporter and all the others have to be quiet. Once the question is answered, the process starts all over. With more and more reporters in the press conference, the lower the chance each one will get to ask a question.

Initially an Ethernet network required each host device have a coaxial cable connection to a backbone coaxial cable (called a "bus"). As more devices were added, the more expensive cable that was needed and connecting coaxial cable to the bus was time-consuming and error prone. To alleviate this problem, a network device called a "hub" was introduced. It allowed many Ethernet devices to plug into the hub, while the hub had only one connection to the Ethernet network. It was cost efficient because it reduced the amount of cabling required in the physical Ethernet network. However, it made the collision and congestion problem much worse.

There had to be a way such that each host can get some network bandwidth. This was accomplished using an Ethernet "switch." Like a hub, a switch is an Ethernet network device. Even though packets that are being processed by the switch have information in the packet from all seven OSI layers, the switch can only see, or have access to, the Data Link Header (DH). In contrast to a hub, a switch provides a dedicated port connection to each host (Fig. 2.12). Therefore, there is no battle for the host to access the Ethernet network.

Also, a switch has buffers, or memory. If a host cannot access the Ethernet network, the switch will buffer, or store, the host's frames until the bandwidth is available for the frames to be transmitted.

2.4.6.2   Switch and Router Interaction

To summarize the interaction between network routers and switches, Ethernet switches enable communication between host devices on a LAN. If a connection to a network such as the Internet is required, the packet (remember, there is a packet with information at all seven OSI layers always present, but the switch can only see the first two) will be forwarded to a router over one or a few high bandwidth links (Fig. 2.13) The router will then look at Layer 3 at the destination IP address and then consult its route table to determine the best next hop route for the packet. Each router in line will do the same until the packet reaches its destination.

Fig. 2.12 Switch OSI layers

Data
Physical

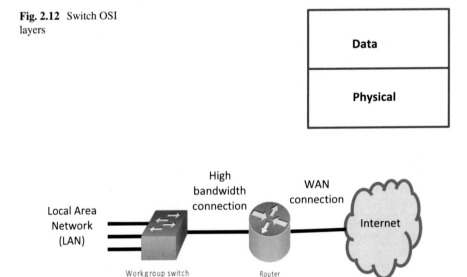

Fig. 2.13 Switch and router interaction

2.4.7 Layer 1: Physical Protocols

The Physical layer is responsible for interfacing with many different media types, for example, light for optical and RF for cellular network. Cloud engineers need not concern themselves with the complexity of this layer.

2.4.8 Putting It All Together

The diagram below illustrates a typical scenario where the customer's hosts communicate locally with each other as well as remotely with applications running on a server in the public cloud provider's data center. The network connection is the Internet.

The box around the customer and the public cloud provider represents what equipment is present at each physical location.

The OSI seven layers are resident in the workstation host of the customer and the server of the public cloud provider. An Ethernet switch connects to the workstation of the customer and the server of the public cloud provider. However, realize that in a real-world example, the switch in the customer network will connect to numerous workstations. Similarly, the switch in the public cloud provider's data center will likely connect to multiple servers. The switches provider for local communication between the host level devices on both ends.

If a customer host needs to connect with an application running in the public cloud provider's data center, it must access the local router that has access to the Internet. The diagram shows only one router in the Internet when in reality there are many routers between the customer and the public cloud provider. Packets are routed across the Internet based on the dynamic routing protocol in use such as least number of hops or shortest path.

2.4.8.1 Packet Walk

We will now see how a packet sequentially "walks" from a host workstation to the remote server (Fig. 2.14). As the packet leaves each device, it is transmitted as bits that represent the created packets. The state of the packet with the OSI layer overhead is presented at each step. Notice the Layer 4 through 7 "A P S T" overhead is present in the packet across the entire network but is not visible by the Layer 2 switches and Layer 3 routers, only the hosts on each end can access all seven layers.

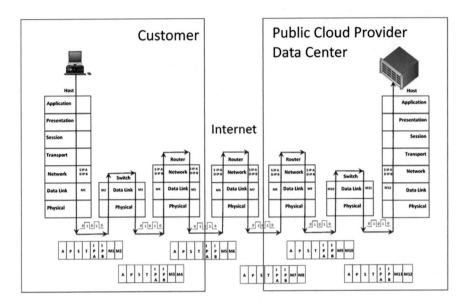

Fig. 2.14 Packet walk

2.4.8.2 IP Addresses

Communication between the customer and the public cloud provider's data center is accomplished using the IP addresses of the source workstation and the destination server. In this case, only communication from the workstation to the server is presented. In reality, there is often return communication as well. "S IP A" is the source IP address of the workstation and "D IP B" is the destination IP address of the server. Notice that the source and destination IP addresses do not change as the packet walks across the network. Also realize that the switches cannot see the IP addresses while each router does.

2.4.8.3 MAC Addresses

As the source and destination IP addresses do not change as a packet walks across the network, MAC addresses change with each hop.

Both switches and router interfaces operate at Layer 2; hence, both devices see MAC addresses in the packet. However, as the packet leaves a switch or router, it places the source MAC address of the devices outgoing NIC as the source MAC address and the destination MAC address of the next hop device's NIC. As the packet walks across the network, the source and destination MAC addresses change with each hop.

2.4.9 The Next Step

At this point, you have learned about networking as defined by the seven-layer OSI model and the network protocols most important in the cloud that implement those functions in hosts, switches, and routers. You have also "walked" a packet across a simplified network connection between a customer and their public cloud provider.

Alternatively, you may decide you have had enough knowledge and want to stop learning. With what you have learned so far, you can jump ahead to the labs and, if you follow the directions exactly, will be successful in completing them. However, you will fall short of the knowledge and you need to thoroughly understand the labs as well to have the background to seriously consider a career in the cloud industry.

You can think of what we have done so far as entering the wide part of a funnel. To really understand networking and cloud operation requires a deeper understanding, which requires applying some basic mathematics. Therefore, if you continue on, we will begin to narrow the funnel.

The material that follows is challenging but can be learned with focus and discipline. If you continue on, there will be times of frustration and exhaustion. But, if you persevere, you will be rewarded with a challenging and financially lucrative career in a relatively short period of time, far quicker than spending 4 years in a college or university. Just remember, they do not pay you the big bucks for doing easy things!

So, fearlessly we venture on! We will start by learning different numbering systems used in networking.

2.5 Summary

Networks have evolved from those carrying just voice telephone calls to those carrying primarily packets of data. For many years, the AT&T telephone network was a monopoly but was broken up to encourage competition in the industry. Simultaneously, the Internet rapidly became a vital communication channel. The Open System Interconnection (OSI) seven-layer model was developed to simplify how communication occurs across networks, as a guide for the standardized development of network devices, and to allow seamless communications between diverse systems.

This chapter introduced the OSI seven-layer model and identified the protocols that belong in each layer. The function and operation of several protocols were explained and how they cooperate to connect packets across a TCP/IP network examined. Details of IP and MAC addresses were covered. Switch and router functions and layer operations are detailed as well as how packets "walk" from host-to-host across the network are illustrated.

Homework Problems and Questions

1.1 Describe step-by-step how a packet "walks" across a TCP/IP network from host to host.

1.2 Discuss the use of IP and Ethernet addressing and how the interact in a TCP/IP network.

1.3 List and describe the operation of the following protocols:

TCP

UDP

DNS

DHCP

1.4 Compare and contrast the functions and operation of switches and routers.

1.5 Describe how packet routing occurs in a TCP/IP network.

1.6 Describe the process of encapsulation and discuss its operation.

1.7 List and discuss each of the seven layers of the OSI model.

1.8 Discuss the advantages of encapsulation using the OSI model.

Bibliography

500-Black. (n.d.). Retrieved from oldphonenetworks.com: https://oldphoneworks.com/products/black-model-1

About MEF. (2023). Retrieved from MEF: https://www.mef.net/

Carterfone, ATT and the FCC 1948–1967. (n.d.). Retrieved from History of Computer Communications: https://historyofcomputercommunications.info/section/3.2/carterfone,-att-and-the-fcc-1948-1967/

DNS Tutorial part 1 – DNS basics. (n.d.). Retrieved from DNS Monitor: https://dnsmonitor.com/dns-tutorial-1-the-basics/

Metcalfe, B. (2013, December). Metcalfe's law after 40 years of Ethernet. *Computer, 46*(12), 26–31. https://doi.org/10.1109/MC.2013.374

National Archives. (2020, May 6). *Telephone and light patent drawings.* Retrieved from National Archives: https://www.archives.gov/education/lessons/telephone-light-patents

Open Systems Interconnection-Model and Notation. (1994, July). Retrieved from ITU-T: file:///C:/Users/01415036/Downloads/T-REC-X.200-199407-I!!PDF-E%20(1).pdf

Running pass on. (2023). Retrieved from iStock: https://www.istockphoto.com/photo/running-pass-on-stadium-gm509412492-85761393?phrase=relay%20race

The Evolution of Telephone Cable. (n.d.). Retrieved from Copper Development Corporation, Inc.: https://www.copper.org/applications/telecomm/consumer/evolution.html

Turner, S. (n.d.). *Theodore V. Vail.* Retrieved from Telecommunications History Group: https://www.telcomhistory.org/resources/online-exhibits/heroes-in-telecom-history/theodore-n-vail/

Chapter 3
Numbering Systems and Subnetting

3.1 Numbering Systems

Machines communicate using digital bits or "ones" and "zeros." Of course, people do not naturally communicate that way. Therefore, to more easily understand what the machines say, various number systems are used.

A cloud engineer must be able to understand data coded using these numbering systems to work beyond an introductory level in the cloud. This chapter identifies and explains the number systems used in networks and the cloud. Learning and applying these numbering systems at first will probably be very confusing. However, remember you are learning a new language. It isn't Spanish or French, but the language of machines. It will take some dedication and concentration, but the average person can master these concepts. Don't get discouraged. Put it aside and try again later. Success will result in the next big step toward a career in the cloud!

First, we will review a numbering system we all recognize.

3.1.1 *Base$_{10}$ Numbering System*

We are all familiar with the base$_{10}$ ("base ten") numbering system. It uses the numbers zero through nine. We were taught in grade school how it works and we use it every day without much thought. For review, we use the illustration of a base$_{10}$ numbering with four places (Table 3.1):

Each of the four places can only be a number between "0" and "9." We will use the number sequence of "2963."

As you recall, the rightmost place is the "ones" or 10^0 (ten-to-the-zero power) place. Numbers in this place can be between "0" and "9." In this case, the place's **value** is $3 \times 1 = 3$.

© The Author(s), under exclusive license to Springer Nature Switzerland AG 2024
M. S. Kingsley, *Cloud Technologies and Services*, Textbooks in
Telecommunication Engineering, https://doi.org/10.1007/978-3-031-33669-0_3

Table 3.1 Base$_{10}$ numbering

2	9	6	3
"Thousands"	"Hundreds"	"Tens"	"Ones"
10^3	10^2	10^1	10^0

The second from the right place is the "tens" or 10^1 (ten-to-the-first power or 10×1) place. Again, numbers in this place can be between "0" to "9." However, any number in this place is multiplied by 10. In this case, the place's *value* is $6 \times 10 = 60$.

The third from the right place is the "hundreds" or 10^2 (ten-to-the-second power or 10×10) place. Again, numbers in this place can be between "0" and "9." However, the number in this place is multiplied by 100, so this place's *value* is $9 \times 100 = 900$.

Finally, the first place from the left is the "thousands" or 10^3 (ten-to-the-third power or $10 \times 10 \times 10$) place. Again, numbers can be between "0" and "9." In this case, the place's *value* is $2 \times 1000 = 2000$.

Notice that as more places are added to the left, the value of the leftmost place increases by a factor of 10. If all the place values above are added together, the value of all four places is 2963.

$$
\begin{aligned}
\left(2 \times 1000\right) &= 2000 \\
+\left(9 \times 100\right) &= 900 \\
+\left(6 \times 10\right) &= 60 \\
+\left(3 \times 1\right) &= 3 \\
\hline
&= 2963
\end{aligned}
$$

This is not an unfamiliar example. However, machines do not communicate using base$_{10}$ numbering. Instead, several other numbering systems are used in the cloud.

3.1.2   Base$_2$ Numbering System

The base$_2$ ("base two") numbering system is the basis of digital systems including computers and networks. It is also called "binary" because there are only two numbers available: "0" and "1." They also represent electrical "on" and "off" in computers.

Similar to how base$_{10}$ numbering was explained, we will now examine base$_2$ numbering (Table 3.2).

The previous concepts are applied to base$_2$ except the places representing single bits that can be either a "0" or a "1." In this case, we will examine the four-bit sequence of "1111."

Table 3.2 Base$_2$ numbering

1	1	1	1
"Eights"	"Fours"	"Twos"	"Ones"
$2^3 = 8$	$2^2 = 4$	$2^1 = 2$	$2^0 = 1$

In base$_2$ numbering, the rightmost place is still the "ones" or 2^0 (two-to-the-zero power) place. (We know that any number to the zero power equals "1.") The number in this place can only be a "0" or "1." In this case, the place's ***value*** is $1 \times 1 = 1$.

The second from the right bit is the "twos" or 2^1 (two-to-the-first power) place. Again, numbers in this bit can only be a "0" or "1." The ***value*** in this place is $1 \times 2 = 2$.

The third from the right place is the "fours" or 2^2 (two-to-the-second power or 2×2) place. Again, the number in this bit can only be a "0" or "1." Therefore, the ***value*** of this place is $1 \times 4 = 4$.

Finally, the first bit on the left is the "eights" or 2^3 (two-to-the-third power or $2 \times 2 \times 2$) place. Again, numbers in this bit can only be a "0" or "1." However, the ***value*** of this place is $1 \times 8 = 8$.

Notice that if another place is added to the left, the value of that place doubles; each place's value is half of the place to its left. This is an important concept that will be applied next.

When all the bit values are added together, the value 1111 is 15 in base$_{10}$ numbering.

$$
\begin{aligned}
(1 \times 8) &= 8 \\
+(1 \times 4) &= 4 \\
+(1 \times 2) &= 2 \\
+(1 \times 1) &= 1 \\
\hline
&\ \ 15
\end{aligned}
$$

Table 3.3 illustrates how to convert between base$_2$ and base$_{10}$ for 0 through 15 with four bits.

3.1.3   IP Addressing

Internet protocol (IP) version 4 (IPv4) addresses (there is a later version 6 to be discussed later) are well recognized. They look similar to **192.168.10.5**. However, very few understand the details of how IP addressing works. Working in the cloud requires you to know more about IP addressing and its operation.

IPv4 addresses are actually 32 binary bits long. For example, 192.168.10.5 is actually:

11000000101010000000101000000101

Table 3.3 Base$_2$ to base$_{10}$ conversion

Base$_2$	Base$_2$ <--> Base$_{10}$	Base$_{10}$
0000	$(0 \times 8) + (0 \times 4) + (0 \times 2) + (0 \times 0)$	0
0001	$(0 \times 8) + (0 \times 4) + (0 \times 2) + (1 \times 1)$	1
0010	$(0 \times 8) + (0 \times 4) + (1 \times 2) + (0 \times 1)$	2
0011	$(0 \times 8) + (0 \times 4) + (1 \times 2) + (1 \times 1)$	3
0100	$(0 \times 8) + (1 \times 4) + (0 \times 2) + (0 \times 1)$	4
0101	$(0 \times 8) + (1 \times 4) + (0 \times 2) + (1 \times 1)$	5
0110	$(0 \times 8) + (1 \times 4) + (1 \times 2) + (0 \times 1)$	6
0111	$(0 \times 8) + (1 \times 4) + (1 \times 2) + (1 \times 1)$	7
1000	$(1 \times 8) + (0 \times 4) + (0 \times 2) + (0 \times 1)$	8
1001	$(1 \times 8) + (0 \times 4) + (0 \times 2) + (1 \times 1)$	9
1010	$(1 \times 8) + (0 \times 4) + (1 \times 2) + (0 \times 1)$	10
1011	$(1 \times 8) + (0 \times 4) + (1 \times 2) + (1 \times 1)$	11
1100	$(1 \times 8) + (1 \times 4) + (0 \times 2) + (0 \times 1)$	12
1101	$(1 \times 8) + (1 \times 4) + (0 \times 2) + (1 \times 1)$	13
1110	$(1 \times 8) + (1 \times 4) + (1 \times 2) + (0 \times 1)$	14
1111	$(1 \times 8) + (1 \times 4) + (1 \times 2) + (1 \times 1)$	15

Table 3.4 Binary value chart

Value	128	64	32	16	8	4	2	1
2^n	2^7	2^6	2^6	2^4	2^3	2^2	2^1	2^0
Bit number	8	7	6	5	4	3	2	1

Obviously, reading and converting the bit stream to base$_{10}$ number are hard. Therefore, we break the 32 bits into four eight-bit chunks, or octets, and separate each by a decimal point. This is called "dotted decimal" notation:

11000000.10101000.00001010.00000101

All right, that is looking a little bit clearer. Now convert each of the octet to a base$_{10}$ number. However, instead of using four bits as we did previously, we use eight. We will refer to each eight bit as an "octet" or "byte." Refer to Table 3.4.

Starting with the farthest right octet (00000101), we see that the base$_{10}$ value is "5" (4 + 1). The second octet from the right (00001010) has a base$_{10}$ value of "10" (8 + 2). The third octet from the left (10101000) has a base$_{10}$ value of "168" (128 + 32 + 8). Finally, look at the first octet on the left (11000000). There are "1" bits in the seventh- and eighth-bit places, so base$_{10}$ value is "192" (128 + 64). Placing all four octet values together results in the dotted decimal representation of the binary IP address of:

192.168.10.5

3.1.4 Base$_{16}$ Numbering System

Another numbering system used in networks and the cloud is base$_{16}$ ("base 16") numbering, also called "hexadecimal" numbering. "Hex" means "six," so in addition to the binary numbers representing "0" through "9," it uses six letters, "A" through "F," to represent the base$_{10}$ numbers of 10 through 15. The table below illustrates how base$_2$ and base$_{10}$ compare to base$_{16}$ numbering (Table 3.5).

3.1.5 Ethernet MAC Addressing

The most recognizable application of hexadecimal numbering is in Ethernet networking. Ethernet MAC addresses are addresses that are "burned in," or permanently assigned to network interface cards. Connection between Ethernet devices is accomplished by using the MAC address of the source network interface card and the MAC address of the destination network interface card. MAC addresses are even longer than IPv4 addresses using 48 bits. For example:

011010001110110011110101001010111101011110100001

We use hexadecimal numbering to make this address more readable. To begin, divide the 48 bits into twelve four-bit chunks separated by dashes:

0110-1000-1110-1100-1111-0101-0010-1011-1101-0111-1010-0001

Table 3.5 Base$_{10}$ to base$_2$ and base$_{10}$ conversion

Base$_2$	Base$_{10}$	Base$_{16}$
0000	0	0
0001	1	1
0010	2	2
0011	3	3
0100	4	4
0101	5	5
0110	6	6
0111	7	7
1000	8	8
1001	9	9
1010	10	A
1011	11	B
1100	12	C
1101	13	D
1110	14	E
1111	15	F

Using the hexadecimal table above results in the MAC address of:

68-EC-F5-2B-D7-A1

I think you will agree that hexadecimal notation, even though still complex, is easier to read than the entire binary address of 48 bits.

3.1.6   IP Version 6 (IPv6)

When the addressing used in the Internet (previously called the "Arpanet") was developed, they did not anticipate the explosive demand placed on it by millions of users. IPv4 provided about four billion IP addresses. All those addresses are now exhausted.

Fortunately, Internet network planners had anticipated the depletion of all IPv4 addresses. The successor to IPv4 is IPv6 (Yes, there was an IPv5, but it was abandoned due to limitations.) While IPv4 used a 32-bit address space, IPv6 uses 128 bits. Remember, as you add a bit in an address space, the number of available addresses doubles. Therefore, the 33rd bit doubled the IPv4 address space of 4 billion addresses to 8 billion; the 34th bit doubled it again and so on through 128 bits. As a result, IPv6 allows for 340 undecillion or 340 trillion, trillion, trillion IP addresses. The actual number of IPv6 addresses is:

340,282,366,920,938,463,463,374,607,431,768,211,456

This is an incomprehensibly large number, large enough that every person on earth could have a billion, billion IP addresses for every second of their life!

While IPv4 uses dotted decimal notation, the IPv6 developers favored hexadecimal addressing. Therefore, going forward, both Ethernet and IPv6 will use hexadecimal addressing. An example of an IPv6 address is:

2603:8080:f80a:f5b6:fc85:5118:700d:8336

Learning and using IPv6 can be difficult because it has many tricks and shortcuts that can be applied. It is possible that an IPv6 address can be correctly presented in several different ways.

There has been a lot of reluctance to embrace IPv6, particularly by public cloud providers. Even though all IPv4 addresses are exhausted from the Internet management authorities, it does not mean IPv4 addresses are not available. There is a large market for purchasing IP addresses on the open market. Amazon Web Services and other public cloud providers have vast IPv4 address inventories. Nonetheless, they are slowly integrating IPv6 capability into their infrastructures. However, for the time being, interacting with public cloud infrastructure is primarily via IPv4. Therefore, we will focus on solutions using IPv4 addressing.

3.2 Address Classes

In the early days of the Internet, the first task was to allocate publicly usable IP addresses. The result was three classes of IPv4 addresses identified by the letters A, B, and C (Reynolds & Postel, 1983). (Two other classes, D and E, were reserved for something called multicasting and experimental use. These are beyond the scope of this discussion.)

- A "**Class A**" address is identified by forcing the *first* bit from the left to be a "0."
- A "**Class B**" address is identified by forcing the first *two* bits from the left to be "10" or 128 in base$_{10}$.
- A "**Class C**" address is identified by forcing the first *three* bits from the left to be "110" or 192 in base$_{10}$.

Therefore, in dotted-decimal format, the address ranges for classes A, B, and C are:

Class A: 0.0.0.0 to 127.255.255.255
Class B: 128.0.0.0 to 191.255.255.255
Class C: 192.0.0.0 to 223.255.255.255

3.2.1 Counting in Dotted-Decimal

Counting in dotted decimal requires some practice. Remember, each dotted-decimal base$_{10}$ number expands into eight binary bits. Each octet ranges from 00000000 to 11111111 in base$_2$ (binary), which is 0 to 255 in base$_{10}$ (Table 3.6).

Therefore, each octet counts from 0 to 255 in base$_{10}$ and 00000000 to 11111111 in base$_2$. For example, the count from 192.0.0.0 continues to 192.0.0.1, 192.0.0.2, and 192.0.0.3 up to 192.0.0.255. The next number is 192.0.1.0.

This may be difficult to visualize. However, if you compare it to counting in base$_{10}$, it is easier, for example, counting from "0" to "9." The next number is "10." As you can see, the last bit decreases to "0," and the next number to the left increases to "1." The same principle applies in dotted-decimal counting except instead of having number "0" through "9," the numbers range from "0" to "255." When the count

Table 3.6 Dotted-decimal to base$_2$ conversion

Dotted-decimal	Binary
0	00000000
1	00000001
2	00000010
3	00000011
↓	↓
255	11111111

reaches "255," the number in that octet goes to "0," and the number in the octet to the left increases by '1."

Therefore, you can now easily see that the last number in a Class A address is 127.255.255.255, and the following number of 128.0.0.0 begins the Class B address range. The last number of the Class B range of 191.255.255.255 advances to 192.0.0.0 to begin the Class C address range. Similarly, 223.255.255.255 is the last number of the Class C address range.

3.3 Networks and Hosts

The next task is to determine which IP addresses in each class are used to define networks and which addresses are assignable to hosts (computers and other devices). This is accomplished by assigning specific octets in the IP address space to networks and hosts.

3.3.1 Class "A" Address Space

The first octet from the left in the Class A address space defines networks (Fig. 3.1). Remember, if the first bit from the left is a "0", the address is a Class A address. Since this bit is already used, there were seven bits left in the first octet. With the total number of possible networks available using seven bits (0000000 to 1111111) being 2^7 numbered from "0" to "127," there were 128 Class A networks. Both networks "0" and "127" are reserved. Therefore, the maximum networks available for use by Class A is 126.

What about hosts? The first octet from the left is used for networks that left the remaining three octets, or 24 bits, to assign to hosts on each network. Twenty-four host bits, or 2^{24} to calculate the total number of hosts, yield over 16 million host addresses available on each of 126 networks.

Class A network addressing was reserved for only the largest of user group such as the US federal government.

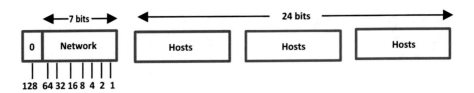

Fig. 3.1 Class A address space

3.3.2   Class "B" Address Space

Very few (if any) user groups require 16 million host addresses for their computers and other devices. The Class B address space accommodates more networks and less hosts for medium-sized user groups (Fig. 3.2).

   Class B uses the first two octets to allocate networks. Recall when the first two bits in the first octet, from left to right, are "10," it identifies the packet as belonging to Class B. Therefore, the Class B address space starts at 128 (1000000 = 128 + 0 + 0 + 0 + 0 + 0 + 0 + 0 = 128). Since two bits (10) have been used to identify the class, there are 14 bits left in the first two octets (6 + 8 = 14) for network numbers. Therefore, the number of Class B networks available is 2^{14}, resulting in 16,834 Class B networks. Similarly, the last two octets, or 16 bits, are reserved for hosts with 2^{16} or more than 65,000 hosts IP addresses for each Class B network.

3.3.3   Class "C" Address Space

To provide IP address space for smaller user groups, Class C addressing could somewhat accommodate (Fig. 3.3). Recall the first three bits of the first octet in Class C is "110," so the Class C space starts with 192 (11000000 = 128 + 64 + 0 + 0 + 0 + 0 + 0 + 0 = 192). Thirty-two bits less the "110" leaves 21 bits in the first three octets to define networks. Therefore, with 2^{21}, there are over two million networks available in Class C. The remaining eight bits in the last octet are available to assign to hosts. Only 2^8 bits, or 256 bits, are allowed for hosts for each of the over two million defined Class C networks.

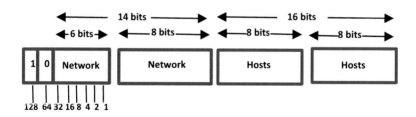

Fig. 3.2 Class B address space

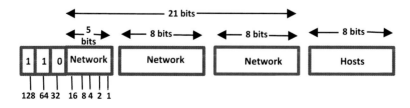

Fig. 3.3 Class "C" address space

Table 3.7 Class C fourth octet

Description	4th Octet Bits							
Bit number in 4th octet	1st	2nd	3rd	4th	5th	6th	7th	8th
Bit number in IP address	25th	26th	27th	28th	29th	30th	31st	32nd
0	0	0	0	0	0	0	0	0
255	1	1	1	1	1	1	1	1

The chart below illustrates the last octet bits (1–8) as well as highlights that these bits are bits 25–32 of the IP address. It also shows the octet can have a value of 00000000 to 11111111 (0–255) (Table 3.7).

3.4 Problems with Classful IP Address Assignment

Assigning IP addresses strictly by classes presents problems. Most noticeable were very few organizations are large enough to consume millions, even thousands, of IP addresses. If they are not used by their assigned owner, they are wasted. For Class A and Class B, the IP addresses that are wasted can be in thousands.

Even if a Class C address space were assigned to a group of users, IP addresses will often be unused and therefore wasted. For instance, even a small company with a few employees may require at most a dozen IP addresses. If IP addresses are assigned strictly by the Class C definition, 256 addresses would have to be assigned to the company. Over 240 IP addresses would be unused and cannot be used by any other company, so they will probably be wasted.

3.4.1 Conserving IP Addresses

Beginning in the 1993, the Internet was accessible to the public. With the addition of the World Wide Web, its popularity grew exponentially. Soon, millions of users were accessing the Internet. It was apparent that if demand continued to grow at that rate, the four billion IPv4 addresses would one day be exhausted. Something had to be done to slow the use of public IP addresses. Several solutions were developed.

3.4.1.1   Private IP Addressing

One of the first solutions was to create private IP addressing spaces. Organizations were assigned public IP addresses to use on the Internet side of routers. However, on the organizations' internal network behind the router, private IP addresses should be used (Fig. 3.4). Private IP addresses make connections inside the organizations' network, but they cannot access the Internet. If an internal network user wants to access the Internet, they will have to be connected to a public IP address using network address translation (NAT). Different users could use the same private IP addresses on their private networks because the routers on the Internet cannot see them behind the company's Internet-facing router. As a result, thousands of public IP addresses were conserved.

Each IP address class designated a range of *private IP addresses* that could be reused as needed (Rekhter et al., 1996):

- **Class A:** 10.0.0.0 – 10.255.255.255
- **Class B:** 172.16.0.0 – 172.31.255.255
- **Class C:** 192.168.0.0 – 192.168.255.255

3.4.1.2   Dynamic Host Configuration Protocol (DHCP)

Another method of decreasing the number of public IP addresses used was by using the Dynamic Host Configuration Protocol (DHCP). To explain how DHCP works, assume there is an Internet service provider (ISP) that has 10,000 customers. It is unlikely that all their customers need to access the Internet at the same time. Assume further that only 30% of their customers on average will access the Internet simultaneously. By only assigning 3000 IP addresses to the ISP and using DHCP to allocate IP addresses as needed, 7000 IP addresses are conserved.

Occasionally, the demand for Internet access may exceed 3000 users. In that case, they would have to wait until someone ended their connection and an IP address could be assigned to them. To ensure this does not happen often, statistical estimation methods to determine the number of IP addresses were needed.

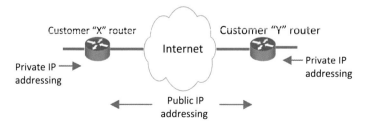

Fig. 3.4 Public and private IP addresses

3.4.1.3 Subnetting Introduction

Subnetting is another way to save dwindling IPv4 addresses. We will investigate subnetting in detail but to do so requires understanding virtual networking. At least until IPv6 is fully implemented, subnetting is a "must-have" skill for the network engineer!

3.5 Welcome to Virtual Networking

The public cloud provider interface to users is implemented in the virtual private cloud (VPC). You can think of a VPC as a box where your compute, database, storage, and many other cloud services you have created are connected via a network (Fig. 3.5). These are software services that physically reside as virtual machines (VMs) on hardware servers in the cloud provider's infrastructure-somewhere. Your compute VM may be in one physical location talking to your database VM is physically hundreds of miles away. From your viewpoint, they appear to be physically close together, while in reality, they probably are not.

In the physical world, networking revolves around routers. Routers are hardware devices that have IP addresses and "route" packets through a network. Network engineers will be confused by the AWS environment because network hardware such as routers and switches are invisible to cloud users. These devices are still there but are managed by the cloud provider, and their details are "abstracted," or hidden, from users.

As a cloud engineer, you will need extensive knowledge of how an IP network operates. You will physically define route tables that determine where your packets are routed in your VPC. The design of the infrastructure including subnets, IP address planning, and implementing network support services such as DHCP, NAT, DNS, and others are also your responsibility. A cloud engineer is a network engineer with virtualization skills.

Fig. 3.5 AWS Virtual Private Cloud (VPC)

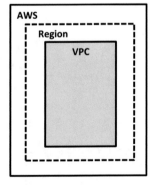

A concern may be that you do not know how to configure, manage, and trouble-shoot your VPC environment. Much of that can be accomplished automatically for you by the cloud provider. This is acceptable for a small cloud infrastructure; however, relying on it for large cloud projects would be asking for disaster.

3.6 Moving Forward

The next step is to dive deeper into VCP networking. Even though the routers are invisible, the VPC network is a typical network. To understand the network operation in the example that follows, we include the routers as if they are visible because VMs have MAC and IPv4 addresses. Staying familiar will make understanding the VPC operation easier. As you will see, all traditional networking concepts are the same in the VPC as they are in any physical network. Actually, the router used in the example could be any Layer 3 device such as compute VM.

We are now ready to learn more about subnetting, a vital concept in network, and therefore VPC, design, and deployment. However, our investigation will focus on using Class C subnetting. Class A and B are briefly discussed at the end of the chapter.

3.7 Subnetting Explained (Finally)

The goal of subnetting is to increase the number of networks and decrease the number of hosts to conserve IP addresses. It is accomplished by manipulating the digits in the IPv4 address space. Recall that in binary numbering, adding a digit to a bit stream doubles the numbers available; taking a digit away from a bit stream will decrease the available numbers by half. This is the principle that is applied in subnetting.

Subnetting will be demonstrated for both classful and classless networks.

3.8 Classful Network Subnetting

When a packet is received by a Layer 3 device, it has to determine where the network and host bits are. Using the original IP addressing scheme, the *class* of the packet was identified in the first bits of the address first octet. Recall if the first bit in the first octet was a "0," it was a Class A address; "10" in the first two bits indicates a Class B address, and "110" in the first three bits indicates a Class C address. This is *classful networking*. The location of the network and the host bits are always the same in each class.

For simplicity, we will focus on Class C to learn subnetting. Classes A and B subnetting require a lot of explanation. Their use will briefly be discussed at the end of the chapter. Details of subnetting Classes A and B are available on the book website.

Recall that Class C is the smallest defined class with 256 host addresses. This is far too many IP addresses for many users. Most of these IP addresses will be wasted. Therefore, the Class C space can be divided into smaller *subnetworks*, or just *subnets*, each with the same number of IP addresses. This is accomplished by "robbing" or borrowing bits from the fourth octet. For every bit borrowed, the number of host addresses is decreased by half.

3.8.1 Subnet Masks

However, classful subnetting caused a problem. The class of an IP address was previously identified by looking at the first few bits of the address space to identify network and host bits that references have now shifted. Which IP addresses belong to whom? The receiving device needs a way to know which bits had changed to properly process the packet. This was accomplished using a *subnet mask*.

A subnet mask is a separate stream of bits, also 32 bits long, from the IP address. A "1" in a bit space of the subnet mask indicated that the corresponding bit space in the IP address is a network bit, and a "0" indicates a host bit. The receiving router now understands where the network and host bits are.

3.8.2 Default Subnet Mask

A "default" subnet mask is the number of bits used to identify the network bits in the original IP address classes.

For a Class A address, the first eight bits identify the network bits. Therefore, the first eight bits of the subnet mask are all 1s. 11111111 base$_2$ results in a base$_{10}$ value of 255. Therefore, the default subnet mask of a Class A IP address is 255.0.0.0 (Fig. 3.6).

A Class B address uses the first two octets, or 16 bits, to identify network bits, so the default subnet mask for Class B is all 1s in the first two octets. 11111111 11111111 base$_2$ results in a base$_{10}$ default subnet mask value of 255.255.0.0 (Fig. 3.7).

For a Class C address, the first three octets determine the network bits, so the default subnet mask is 255.255.255.0 (Fig. 3.8).

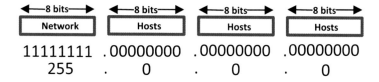

Fig. 3.6   Class A default subnet mask

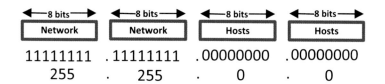

Fig. 3.7   Class B default subnet mask

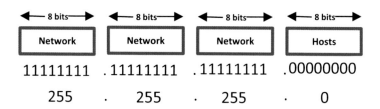

Fig. 3.8   Class C default subnet mask

3.8.3   Custom Subnet Mask

As discussed, IP addresses can be conserved by borrowing bits from the host bits and giving them to the network bits. A give-and-take operation, it simultaneously increases the number of subnetworks and decreases the number of hosts for each of those subnetworks. If you are an ISP, this means you can provide more customers with their own subnets and therefore public IP addresses.

For your VPC, you will be using private IP addressing. Since private IP addresses are plentiful, conserving them is not as crucial as it is for public IP addresses. However, the public cloud provider will not allow you to be sloppy with your planning. If you do not know how to subnet your VPC, the cloud provider can assign IP addresses for you. This is fine for very small cloud implementations, but to be employed as a cloud engineer, you will be expected to be able to effectively assign and manage subnets and host IP address assignments. Failure to plan your cloud infrastructure correctly can have devastating results.

3.8.4 Classless Inter-Domain Routing (CIDR)

For Class C, you can borrow up to six of the eight bits in the fourth octet. If we are configuring network devices (either manually or automatically with DHCP), the subnet mask must be entered in full. However, if you and I are talking, it is inconvenient to say something like "two-fifty-five-two-fifty-five-two-fifty-five dot zero" every time.

Fortunately, we can use a shorthand or abbreviated method to communicate subnet mask information. It is called classless inter-domain routing (CIDR). In the previous example of the Class C address of 192.168.10.5 with a default subnet mask of 255.255.255.0, there are 24 network bits, Therefore, we can write it as 192.168.10.5/24.

As more bits are borrowed, the number after "/" will increase (for a Class C) up to 30 indicating six bits have been borrowed.

3.8.5 Easy Classful Subnetting

Classful subnetting can be easily accomplished using the following eight-step method (Lamle, 2016):

Step 1: What class is the address space?
Step 2: What is the default subnet mask?
Step 3: What are the requirements?
Step 4: What is the custom subnet mask?
Step 5: What is the "magic number (block size)?"
Step 6: Define the subnets.
Step 7: Assign IP addresses to subnets.
Step 8: Assign IP addresses to interfaces.

To illustrate, let us start with a small network. The example network below has four Layer 3 devices. Routers are used to simplify the explanation and are named R1 through R4 (Fig. 3.9). Each router has network interfaces that connect to network links between routers as well as to hosts, the numbers of which are defined in the figure.

Step 1: What class is the address space?

We have been assigned the IPv4 private address space of 192.168.10.0. Notice this is in the Class C private address space.

Step 2: What is the default subnet mask?

Recall that a Class C address uses 24 bits by default to define the network bits. Therefore, the default subnet mask is 255.255.255.0. The CIDR representation is 192.168.10.0/24.

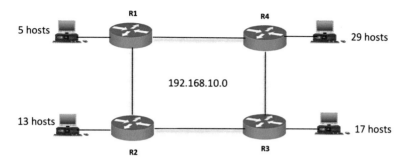

Fig. 3.9 Example network

Step 3: What are the requirements?

Requirements are defined based on the number of subnets or hosts to be supported. We will use subnets. So, how is the number of subnets determined?

Subnet boundaries are Layer 3 interfaces. Each connection between networked device interface(s) is a unique network or subnet. This network needs eight subnets as demonstrated in Fig. 3.10.

Step 4: What is the custom subnet mask?

To get eight subnets, we begin by borrowing one bit from the fourth (host) octet (Fig. 3.11). Doing so will double the subnets to two and simultaneously reduce the number of hosts per subnetwork by half to 128 each. To indicate that the network bits have increased by one, the custom subnet mask in the last octet is .10000000. The value of the first bit is 128, which is also the value of the last octet. Therefore, the full custom subnet mask is 255.255.255.128 or 192.168.10.5/25. However, two subnets are not enough, and 128 IP addresses are also wasteful.

Borrowing a second bit from the fourth octet again doubles the number of subnets from two to four. Each of the four subnets now has 64 IP addresses. The custom subnet mask for the fourth octet is now .11000000 with a value of 128 + 64 = 192. The custom subnet mask is 255.255.255.192 or 192.168.10.0/26. Again, there are not enough subnets and too many IP addresses.

Borrowing a third bit from the fourth octet creates eight subnets with 32 bits each. The custom subnet mask for the fourth octet is now .11100000 with a value of 128 + 64 + 32 = 224. The custom subnet mask is 128 + 64 + 32 = 224. The custom subnet mask is 255.255.255.224 or 192.168.10.5/27. There are enough subnets but still an excess of IP addresses with 32.

Notice that classful subnetting is limited to using subnets from the same column (Table 3.8).

Going one more time, borrowing a fourth bit from the fourth octet returns 16 subnets with 16 IP addresses. Now there is an excess of subnets and too few IP addresses per subnets to accommodate the 17 or 29 IP addresses. Therefore,

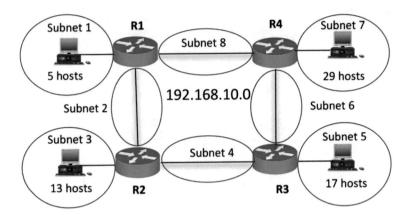

Fig. 3.10   Example network subnets

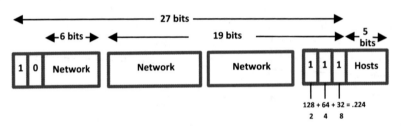

Fig. 3.11   Class C subnetting borrowing three bits

Table 3.8   Binary values chart

Borrowed	1	2	3	4	5	6	
Subnets	2	4	8	16	32	64	
Hosts	128	64	32	16	8	4	
Default	.128	.192	.224	.240	.248	.252	
CIDR	/25	/26	/27	/28	/29	/30	

borrowing three bits from the fourth octet is the best fit. The network space and subnet mask information is:

IP address space: 192.168.10.0
Subnet mask: 255.255.255.224

When the destination router receives the packet, it can determine where to find the network and host bits.

For simplicity, we can talk about this solution by using CIDR:

192.168.10.0/27

Keeping in mind that the Class C address space the fourth octet has reserved for hosts has 256 total host IP addresses available. No matter how many subnets are used, the total number of IP addresses across all subnets cannot exceed 256.

2 subnets × 128 hosts each = 256 IP addresses
4 subnets × 64 hosts each = 256 IP addresses
8 subnets × 32 hosts each = 256 IP addresses
16 subnets × 16 hosts each = 256 IP addresses
32 subnets × 8 hosts each = 256 IP addresses
64 subnets × 4 host each = 256 IP addresses

Step 5: What is the "block size?"

The "block size" is the number of IP addresses assigned to classful subnets. This information was already presented in the previous process. However, a quick way to determine what the block size is for any Class C problem is to subtract the custom subnet mask from 255. In the previous example, $255 - 224 = 32$. Next, divide 256 by 32 ($256/32 = 8$). Therefore, as previously described, there are eight subnets with 32 bits, each of which when added together equals 256.

Step 6: Assign subnets to the network

The next step is to assign subnets to the example network (Table 3.9) starting with .0 and number each subsequent subnet in increments of .32. In this case, the subnet numbers are 192.168.10.0, .32, .64, .96, .128, .160, .192, and .224. Since all subnets are the same size, the order in which they are assigned does not matter.

Notice that regardless of the number of needed IP addresses, a subnet with 32 IP addresses must be used. Therefore, using any less than 32 IP addresses in a subnet

Table 3.9 Subnet assignments

Subnet #	Subnet IP
Subnet 1	.0
Subnet 2	.32
Subnet 3	.64
Subnet 4	.96
Subnet 5	.128
Subnet 6	.160
Subnet 7	.192
Subnet 8	.224

(not considering extra addresses for future growth at this time) will be stranded in a subnet and wasted. Also, notice that if any of the eight subnets requires more than 32 IP addresses, then classful subnetting will not work.

Step 7: Assign IP addresses to subnets

The next step is to assign IP addresses to the subnets from the chart below (Table 3.10):

Step 8: Assign IP addresses to interfaces

Subnets 1, 3, 5, and 7 have connected hosts that need IP addresses. Subnet 1 has only five hosts that need addresses. Remember, the first address in the subnet is the subnet number and the last address is the broadcast address that cannot be assigned to hosts. In addition, the router interface connected to the hosts needs an address as well. A good practice is to assign the first useable IP address to the router interface. Five host addresses plus three reserved IP addresses bring the total addresses required for subnet 1 to eight. That leaves $32 - 8 = 24$ addresses unused. Similarly, subnets 3 and 5 need 16 and 20 IP addresses, respectively, but waste 16 and 12 addresses, respectively. Only subnet 7 with 32 IP addresses required will not waste any IP addresses.

Connections between routers and subnets 2, 4, 6, and 8 require only four IP addresses in their subnet: two IP addresses for the router interfaces on both ends, one for the subnet number, and one for the broadcast address. With 32 addresses in the subnet and only four needed leaves 28 addresses, which are wasted in all four network connecting subnets. They are not needed in this subnet and cannot be used in any other subnet.

Below is the completed classful configuration (Fig. 3.12).

Table 3.10 Subnet IP address assignments

Subnet #	Subnet IP	First useable	Last useable	Broadcast
Subnet 1	.0	.1	.30	.31
Subnet 2	.32	.33	.62	.63
Subnet 3	.64	.65	.94	.95
Subnet 4	.96	.97	.126	.127
Subnet 5	.128	.129	.158	.159
Subnet 6	.160	.161	.190	.191
Subnet 7	.192	.193	.222	.223
Subnet 8	.224	.225	.254	.255

Fig. 3.12 Subnet and interface IP address assignments

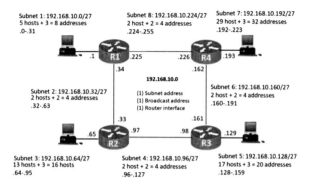

Table 3.11 Summary of classful subnetting

Subnet #	Network	First useable	Last useable	Broadcast	Addresses used	Wasted addresses
Subnet 1	.0	.1	.3	.32	5 + 3 = 8	24
Subnet 2	.32	.33	.60	.61	2 + 2 = 4	28
Subnet 3	.64	.65	.94	.95	13 + 3 = 16	16
Subnet 4	.96	.97	.126	.127	2 + 2 = 4	28
Subnet 5	.128	.129	.158	.159	17 + 3 = 20	12
Subnet 6	.160	.161	.190	.191	2 + 2 = 4	28
Subnet 7	.192	.193	.222	.223	29 + 3 = 32	0
Subnet 8	.224	.225	.254	.255	2 + 2 = 4	28
					92	164
						92 + 164 = 256

3.8.6   *Classful Subnetting Summary*

Classful subnetting is seldom efficient and has many disadvantages. In this example, 164 of the 256 IP addresses are wasted (Table 3.11). The IP addresses used and those wasted equal all the IP addresses available of 256. This network is gridlocked because all the IP addresses are assigned even though many are stranded and unused. This classful design provides only 36% efficiency of IP addresses.

This is only an introduction to subnetting. As you improve your subnetting skills, you will reference the full Class A, B, and C IP address charts. Table 3.12 is the full Class C chart with the previous classful subnets highlighted under the /27 column.

Table 3.12 Class C classful subnet chart using /27

/25	/26	/27	/28	/29	/30
.10000000	.11000000	.11100000	.11110000	.11111000	.11111100
128+0+0+0+0+0+0+0	128+64+0+0+0+0+0+0	128+64+32+0+0+0+0+0	128+64+32+16+0+0+0+0	128+64+32+16+8+0+0+0	128+64+32+16+8+4+0+0
.128	.192	.224	.240	.248	.252
2 subnets	4 subnets	8 subnets	16 subnets	32 subnets	64 subnets
128-2 hosts each	64-2 hosts each	32-2 hosts each	16-2 hosts each	8-2 hosts each	4-2 hosts each
.0 (.1-.127)	.0 (.1-.63)	.0 (.1-.31)	.0 (.1-.15)	.0 (.1-.7)	.0 (.1-.3)
					.4 (.5-.7)
				.8 (.9-.15)	.8 (.9-.11)
					.12 (.13-.15)
			.16 (.17-.31)	.16 (.17-.23)	.16 (.17-.19)
					.20 (.21-.23)
				.24 (.25-.31)	.24 (.25-.26)
					.28 (.29-.31)
		.32 (.33-.63)	.32 (.33-.47)	.32 (.33-.39)	.32 (.33-.34)
					.36 (.37-.39)
				.40 (.41-.47)	.40 (.41-.42)
					.44 (.45-.47)
			.48 (.49-.63)	.48 (.49-.55)	.48 (.49-.51)
					.52 (.53-.54)
				.56 (.57-.63)	.56 (.57-.59)
					.60 (.61-.62)
	.64 (.65-.127)	.64 (.65-.95)	.64 (.65-.79)	.64 (.65-.71)	.64 (.65-.67)
					.68 (.69-.71)
				.72 (.73-.79)	.72 (.73-.75)
					.76 (.77-.79)
			.80 (.81-.95)	.80 (.81-.87)	.80 (.81-.83)
					.84 (.85-.86)
				.88 (.89-.95)	.88 (.89-.91)
					.92 (.93-.94)
		.96 (.97-.127)	.96 (.97-.111)	.96 (.97-.103)	.96 (.97-.99)
					.100 (.101-.103)
				.104 (.105-.111)	.104 (.105-.107)
					.108 (.109-.111)
			.112 (.113-.127)	.112 (.113-.119)	.112 (.113-.115)
					.116 (.117-.119)
				.120 (.121-.127)	.120 (.121-.123)
					.124 (.125-.127)
.128 (.129-.255)	.128 (.129-.191)	.128 (.129-.159)	.128 (.129-.143)	.128 (.129-.135)	.128 (.129-.131)
					.132 (.133-.135)
				.136 (.137-.143)	.136 (.137-.139)
					.140 (.141-.143)
			.144 (.145-.159)	.144 (.145-.151)	.144 (.145-.147)
					.148 (.149-.151)
				.152 (.153-.159)	.152 (.153-.155)
					.156 (.157-.159)
		.160 (.161-.191)	.160 (.161-.175)	.160 (.161-.167)	.160 (.161-.163)
					.164 (.165-.167)
				.168 (.169-.175)	.168 (.169-.171)
					.172 (.173-.175)
			.176 (.177-.191)	.176 (.177-.183)	.176 (.177-.179)
					.180 (.181-.183)
				.184 (.185-.191)	.184 (.185-.187)
					.188 (.189-.191)
	.192 (.193-.255)	.192 (.193-.223)	.192 (.193-.207)	.192 (.193-.199)	.192 (.193-.195)
					.196 (.197-.199)
				.200 (.201-.207)	.200 (.201-.203)
					.204 (.205-.207)
			.208 (.209-.223)	.208 (.209-.215)	.208 (.209-.211)
					.212 (.213-.215)
				.216 (.217-.223)	.216 (.217-.219)
					.220 (.221-.223)
		.224 (.225-.255)	.224 (.225-.239)	.224 (.225-.231)	.224 (.225-.227)
					.228 (.229-.231)
				.232 (.233-.239)	.232 (.233-.235)
					.236 (.237-.239)
			.240 (.241-.255)	.240 (.241-.247)	.240 (.241-.243)
					.244 (.245-.247)
				.248 (.249-.255)	.248 (.249-.251)
					.252 (.253-.255)

3.8.7 Problems with Classful Subnetting

This example illustrates some of the problems with classful subnetting:

1. The subnets have consumed all of the possible IP addresses. Expanding the network is not possible.
2. Between the used and wasted IP addresses, all 256 addresses have been taken. Even though there are IP addresses left in each subnet that can be used if other hosts are attached to that subnet, they cannot be used in any other subnet.
3. Increasing the number of hosts is not possible in one subnet. Subnet 7 has used all 32 IP addresses and cannot add any more hosts.

Fortunately, many classful subnetting limitations can be overcome using *classless* subnetting.

3.9 Classless Subnetting

Looking at the previously used classful subnetting example and the related Class C subnet chart, you may ask the question "Why do I have to stay in just one column of the chart?" In other words, "Why can't I color outside of the lines?" The answer is you can. In doing so, we eliminate many of the disadvantages of classful subnetting. We will examine several problems with classful subnetting below and see how classless subnetting provides solutions to those problems.

3.9.1 Problem 1: Need to Conserve IP Addresses

In classless subnetting, the number of IP addresses needed is considered first (Fig. 3.4). Starting with the subnet with the most IP addresses, the best-fit subnet is selected. In this case, it is subnet 7 that needs 32 total IP addresses (Recall that we added three IP addresses to the 29 host addresses for subnet number, broadcast, and router interface.) The .0 subnet is chosen using the /27 column as before. Also, as before, all the IP addresses are used in this subnet, and no addresses are wasted (Table 3.13).

The subnet with the second most IP addresses is subnet 5 with 20 IP addresses. In this case, the .32 subnet of /27 is selected because a /28 only has 16 IP addresses, which are not enough. A /26 subnet would allow 64 IP addresses that would be even more wasteful than the /27 subnet. Therefore, the best solution is to again use the /27 subnet that still wastes 12 IP addresses.

The subnet with the next most needed IP addresses is subnet 3 requiring 16 IP addresses. With classful subnetting, the subnet was forced to use a /27 wasting 16 IP addresses. However, with classless subnetting, a /28 subnet starting with .64 can be used. No IP addresses are wasted.

Table 3.13 Classless subnetting with IP addresses use and wasted

Subnet	Subnet #	First useable address	Last useable address	Broadcast address	Used addresses	Wasted addresses
Subnet 7	.0	.1	.30	.31	29 + 3 = 32	0
Subnet 5	.32	.33	.62	.63	17 + 3 = 20	12
Subnet 3	.64	.65	.78	.79	13 + 3 = 16	0
Subnet 1	.80	.81	.86	.87	5 + 3 = 8	0
Subnet 2	.88	.89	.90	.91	2 + 2 =4	0
Subnet 4	.92	.93	.94	.95	2 + 2 =4	0
Subnet 6	.96	.97	.98	.99	2 + 2 =4	0
Subnet 8	.100	.101	.102	.103	2 + 2 =4	0
					92	12

Table 3.14 Using multiple subnet sizes

Borrowed	1	2	3	4	5	6	
Subnets	2	4	8	16	32	64	
Hosts	128	64	32	16	8	4	
Default	.128	.192	.224	.240	.248	.252	
CIDR	/25	/26	/27	/28	/29	/30	

The next subnet to consider is subnet 1, which needs eight IP addresses. Looking at the Class C chart, a /29 subnet starting with .80 is an exact match with eight IP addresses per subnet. Again, no IP addresses are wasted.

Now consider the 2, 4, 6, and 8 subnets. These are direct connections between network devices. They only require IP addresses for the network connections, the subnet number, and the broadcast for a total of four addresses. With classful subnetting, each subnet wastes 28 IP addresses. Applying classful subnetting using /30, each subnet now uses only four IP addresses each from subnets .88, .92, .96, and .100.

From Table 3.14, you can observe that we have minimized wasted IP addresses by using variably sized subnets that meet the specific needs of each subnet.

Just as we did for classful subnetting, we can expand the chart above to observe the effect of classless subnetting on a full Class C subnet chart (Table 3.15). Note how much more dffiecint classless subnetting is as compared to classful.

Now we can also observe the result of classless subnetting in the fully annotated diagram in Fig. 3.13.

Table 3.15 Full Class C subnet chart

/25	/26	/27	/28	/29	/30
.10000000	.11000000	.11100000	.11110000	.11111000	.11111100
128+0+0+0+0+0+0+0	128+64+0+0+0+0+0+0	128+64+32+0+0+0+0+0	128+64+32+16+0+0+0+0	128+64+32+16+8+0+0+0	128+64+32+16+8+4+0+0
.128	.192	.224	.240	.248	.252
2 subnets	4 subnets	8 subnets	16 subnets	32 subnets	64 subnets
128-2 hosts each	64-2 hosts each	32-2 hosts each	16-2 hosts each	8-2 hosts each	4-2 hosts each
.0 (.1-.127)	.0 (.1-.63)	.0 (.1-.31)	.0 (.1-.15)	.0 (.1-.7)	.0 (.1-.3)
					.4 (.5-.7)
				.8 (.9-.15)	.8 (.9-.11)
					.12 (.13-.15)
			.16 (.17-.31)	.16 (.17-.23)	.16 (.17-.19)
					.20 (.21-.23)
				.24 (.25-.31)	.24 (.25-.26)
					.28 (.29-.31)
		.32 (.33-.63)	.32 (.33-.47)	.32 (.33-.39)	.32 (.33-.34)
					.36 (.37-.39)
				.40 (.41-.47)	.40 (.41-.42)
					.44 (.45-.47)
			.48 (.49-.63)	.48 (.49-.55)	.48 (.49-.51)
					.52 (.53-.54)
				.56 (.57-.63)	.56 (.57-.59)
					.60 (.61-.62)
	.64 (.65-.127)	.64 (.65-.95)	.64 (.65-.79)	.64 (.65-.71)	.64 (.65-.67)
					.68 (.69-.71)
				.72 (.73-.79)	.72 (.73-.75)
					.76 (.77-.79)
			.80 (.81-.95)	.80 (.81-.87)	.80 (.81-.83)
					.84 (.85-.86)
				.88 (.89-.95)	.88 (.89-.91)
					.92 (.93-.94)
		.96 (.97-.127)	.96 (.97-.111)	.96 (.97-.103)	.96 (.97-.99)
					.100 (.101-.103)
				.104 (.105-.111)	.104 (.105-.107)
					.108 (.109-.111)
			.112 (.113-.127)	.112 (.113-.119)	.112 (.113-.115)
					.116 (.117-.119)
				.120 (.121-.127)	.120 (.121-.123)
					.124 (.125-.127)
.128 (.129-.255)	.128 (.129-.191)	.128 (.129-.159)	.128 (.129-.143)	.128 (.129-.135)	.128 (.129-.131)
					.132 (.133-.135)
				.136 (.137-.143)	.136 (.137-.139)
					.140 (.141-.143)
			.144 (.145-.159)	.144 (.145-.151)	.144 (.145-.147)
					.148 (.149-.151)
				.152 (.153-.159)	.152 (.153-.155)
					.156 (.157-.159)
		.160 (.161-.191)	.160 (.161-.175)	.160 (.161-.167)	.160 (.161-.163)
					.164 (.165-.167)
				.168 (.169-.175)	.168 (.169-.171)
					.172 (.173-.175)
			.176 (.177-.191)	.176 (.177-.183)	.176 (.177-.179)
					.180 (.181-.183)
				.184 (.185-.191)	.184 (.185-.187)
					.188 (.189-.191)
	.192 (.193-.255)	.192 (.193-.223)	.192 (.193-.207)	.192 (.193-.199)	.192 (.193-.195)
					.196 (.197-.199)
				.200 (.201-.207)	.200 (.201-.203)
					.204 (.205-.207)
			.208 (.209-.223)	.208 (.209-.215)	.208 (.209-.211)
					.212 (.213-.215)
				.216 (.217-.223)	.216 (.217-.219)
					.220 (.221-.223)
		.224 (.225-.255)	.224 (.225-.239)	.224 (.225-.231)	.224 (.225-.227)
					.228 (.229-.231)
				.232 (.233-.239)	.232 (.233-.235)
					.236 (.237-.239)
			.240 (.241-.255)	.240 (.241-.247)	.240 (.241-.243)
					.244 (.245-.247)
				.248 (.249-.255)	.248 (.249-.251)
					.252 (.253-.255)

Fig. 3.13 Classless
subnetting results

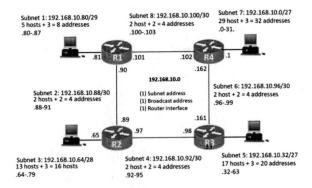

3.9.2   Problem 2: Need for Larger Subnets

The advantages of using classless subnetting is evident when considering network growth possibilities.

Assume that a group of 120 hosts needs to be added to the network, and they must be in the same subnet. With classful subnetting, the largest subnet possible is a /27 subnet with 32 hosts each.

The solution, again, is to use classless subnetting. Continuing with our classless subnetting example, notice there is now an entire /25 address block available with a total of 128 total IP addresses, of which 125 are useable (Table 3.16). However, to use it, a router must be added as well as a link to the network. The link is assigned to the /30 subnet with four IP addresses starting at .104. All 120 hosts can now be connected to a /25 subnet. Additional number of IP addresses used is 127 (123 + 4).

The network, subnets, and IP address assignments now appear as shown in Fig. 3.14.

3.9.3   Spare IP Addresses

The previous example showed the efficiency of using classless subnetting to save IP addresses. In practice, the engineer should choose subnets that have room for growth in case hosts need to be added to a subnet.

3.10   IP Address Planning

Remember, conserving public IP addresses is critical in public networks, whereas in the private network space, there are plenty of private IP addresses that can be reused over and over again. However, good IP address planning is still vital even when

Table 3.16 Adding larger subnet to classless Class C chart

/25 .10000000 128+0+0+0+0+0+0+0 .128 2 subnets 128-2 hosts each	/26 .11000000 128+64+0+0+0+0+0+0 .192 4 subnets 64-2 hosts each	/27 .11100000 128+64+32+0+0+0+0+0 .224 8 subnets 32-2 hosts each	/28 .11110000 128+64+32+16+0+0+0+0 .240 16 subnets 16-2 hosts each	/29 .11111000 128+64+32+16+8+0+0+0 .248 32 subnets 8-2 hosts each	/30 .11111100 128+64+32+16+8+4+0+0 .252 64 subnets 4-2 hosts each
.0 (.1-.127)	.0 (.1-.63)	.0 (.1-.31)	.0 (.1-.15)	.0 (.1-.7)	.0 (.1-.3)
					.4 (.5-.7)
				.8 (.9-.15)	.8 (.9-.11)
					.12 (.13-.15)
			.16 (.17-.31)	.16 (.17-.23)	.16 (.17-.19)
					.20 (.21-.23)
				.24 (.25-.31)	.24 (.25-.26)
					.28 (.29-.31)
		.32 (.33-.63)	.32 (.33-.47)	.32 (.33-.39)	.32 (.33-.34)
					.36 (.37-.39)
				.40 (.41-.47)	.40 (.41-.42)
					.44 (.45-.47)
			.48 (.49-.63)	.48 (.49-.55)	.48 (.49-.51)
					.52 (.53-.54)
				.56 (.57-.63)	.56 (.57-.59)
					.60 (.61-.62)
	.64 (.65-.127)	.64 (.65-.95)	.64 (.65-.79)	.64 (.65-.71)	.64 (.65-.67)
					.68 (.69-.71)
				.72 (.73-.79)	.72 (.73-.75)
					.76 (.77-.79)
			.80 (.81-.95)	.80 (.81-.87)	.80 (.81-.83)
					.84 (.85-.86)
				.88 (.89-.95)	.88 (.89-.91)
					.92 (.93-.94)
		.96 (.97-.127)	.96 (.97-.111)	.96 (.97-.103)	.96 (.97-.99)
					.100 (.101-.103)
				.104 (.105-.111)	.104 (.105-.107)
					.108 (.109-.111)
			.112 (.113-.127)	.112 (.113-.119)	.112 (.113-.115)
					.116 (.117-.119)
				.120 (.121-.127)	.120 (.121-.123)
					.124 (.125-.127)
.128 (.129-.255)	.128 (.129-.191)	.128 (.129-.159)	.128 (.129-.143)	.128 (.129-.135)	.128 (.129-.131)
					.132 (.133-.135)
				.136 (.137-.143)	.136 (.137-.139)
					.140 (.141-.143)
			.144 (.145-.159)	.144 (.145-.151)	.144 (.145-.147)
					.148 (.149-.151)
				.152 (.153-.159)	.152 (.153-.155)
					.156 (.157-.159)
		.160 (.161-.191)	.160 (.161-.175)	.160 (.161-.167)	.160 (.161-.163)
					.164 (.165-.167)
				.168 (.169-.175)	.168 (.169-.171)
					.172 (.173-.175)
			.176 (.177-.191)	.176 (.177-.183)	.176 (.177-.179)
					.180 (.181-.183)
				.184 (.185-.191)	.184 (.185-.187)
					.188 (.189-.191)
	.192 (.193-.255)	.192 (.193-.223)	.192 (.193-.207)	.192 (.193-.199)	.192 (.193-.195)
					.196 (.197-.199)
				.200 (.201-.207)	.200 (.201-.203)
					.204 (.205-.207)
			.208 (.209-.223)	.208 (.209-.215)	.208 (.209-.211)
					.212 (.213-.215)
				.216 (.217-.223)	.216 (.217-.219)
					.220 (.221-.223)
		.224 (.225-.255)	.224 (.225-.239)	.224 (.225-.231)	.224 (.225-.227)
					.228 (.229-.231)
				.232 (.233-.239)	.232 (.233-.235)
					.236 (.237-.239)
			.240 (.241-.255)	.240 (.241-.247)	.240 (.241-.243)
					.244 (.245-.247)
				.248 (.249-.255)	.248 (.249-.251)
					.252 (.253-.255)

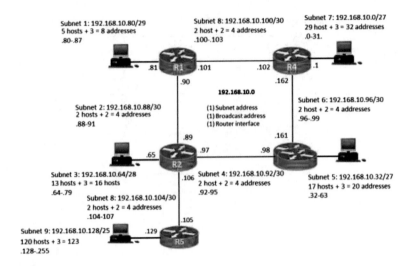

Fig. 3.14 Network

using private addressing. If not done carefully, especially for larger and more complex cloud infrastructures, several problems can occur including delayed application response time and infrastructure failure. Another scenario is where you have to
renumber hundreds of IP addresses in your VPC which would require extensive
downtime and introduce other problems in the environment. All of these circumstances will make you very unpopular, perhaps even get you fired!

Subnetting inside a public cloud provider can be straightforward for small implementations. If you do not want to assign IP addresses to cloud resources, the AWS
configuration wizard will do so for you automatically. However, as your cloud infrastructure grows, so will the knowledge and skill you need to properly design your
infrastructure including effective IP address planning. Having a deeper understanding of networking and subnetting is required to avoid serious problems. However,
there are a few tips that may simplify matters.

3.10.1   Common Private Network Subnetting Schemes

Except for very simple cloud projects, using Class C addressing for your cloud
infrastructure is not recommended due to the limited address space. In the private
address space, there are plenty of IP addresses in Classes A and B. However, proper
IP address planning is vital.

3.10.2 Class B Addressing

Using Class B addressing is straightforward; for example, using 172.16.0.0/24 space. Recall that Class B has a default subnet mask of /16. Using the /24 custom subnet mask results in 256 subnets with 256 host IP addresses for each subnet. Using 172.16.0.0/24 is quick and easy. It provides plenty of subnets and host addresses. This scheme allows plenty of room for growth. Remember, there is no need to worry about wasting private IP addresses!

3.10.3 Class A Addressing

Similar to the Class B example above, Class A addressing can also be easy to use. The Class A address space has a default subnet mask of /8. Using the 10.0.0.0/16 subnet mask provides 256 subnets with over 65,000 hosts per subnet.

3.10.4 Subnet Calculators

It is important that you know how to subnet not only to properly design your cloud infrastructure but also to communicate with colleagues about how and why you designed a VPC the way you did. You will need to talk to them using the language of your vocation. If you cannot do so, you will probably not get a cloud engineering position to begin with. Cloud engineers who can't engineer are called cloud administrators.

You may already know that subnetting calculators are readily available; you can use them. However, a subnet calculator may not be accurate (happened to me!). You need to know how to subnet so that you can detect if there are errors in the IP plan design.

I recommend that when designing an IP subnetting and addressing plan, at first do from the high level or completely manually and use the subnet calculator to check your results. You will be very familiar with your infrastructure and will be able to easily and thoroughly defend your design as well as be more prepared to plan for future expansion.

Subnet calculators range from simplistic to fairly detailed outputs. An example of a subnet calculator is shown in Fig. 3.15.

Bottom line: If you decide to be a cloud engineer after using this book, learn subnetting and other network technologies presented in it in detail. For now, though, get familiar with the topics to follow.

Fig. 3.15 Subnet
calculator example

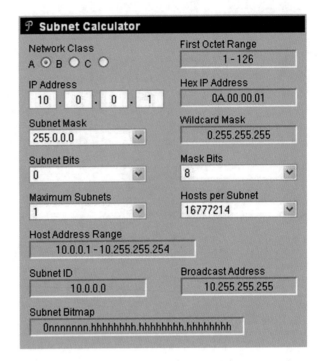

3.11   Summary

This chapter explained the need for and application of both classful and classless
IPv4 subnetting in a VPC. Particular emphasis was given to the importance of
proper network planning of cloud infrastructures and the necessity that cloud engineers be skilled in creating a VPC with proper network design and functionality.

Homework Problems and Questions

Assign subnets and IP addresses to the network below using the 192.168.10.0
address space.

1.1

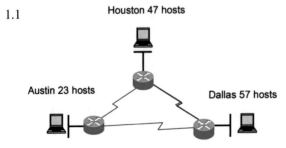

1.2

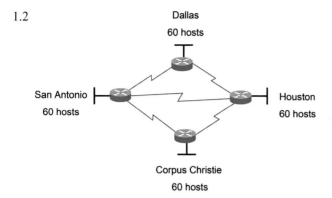

Dallas
60 hosts

San Antonio
60 hosts

Houston
60 hosts

Corpus Christie
60 hosts

1.3   Each top router has 29 hosts

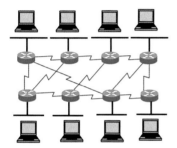

Each bottom router has 13 hosts

1.4 Using an online subnet calculator, complete exercise 1.1–1.3. Compare to your manual results for each of the three networks.

1.5 Convert $base_{10}$ 134 to $base_2$ and $base_{16}$.

1.6 Convert $base_{16}$ 123F to $base_{10}$ and $base_2$.

1.7 Convert 10101010 $base_2$ to $base_{10}$ and $base_{16}$.

1.8 Identify the Class A, B, and C private IPv4 address ranges and discuss the purpose and advantages of private IP addresses.

1.9 Identify the network and host allocations of Class A, B, and C IPv4 addresses.

1.10 Discuss the purpose and use of a subnet mask.

References

Lamle, T. (2016). *CCNA routing and switching study guide*. Cisco Press.

Rekhter, Y., et al. (1996). [Online]. Available at: https://www.rfc-editor.org/rfc/rfc1918

Reynolds, J., & Postel, J. (1983). *Request for comment 870: Assigned numbers* [Online]. Available at: https://datatracker.ietf.org/doc/html/rfc870

Chapter 4
Bringing It All Together

4.1 Cloud Networking in a VPC

All the components of your infrastructure and its network connectivity are contained in a virtual private cloud (VPC) that you define. A VPC is a small part of the public cloud provider's resources that you have access to. However, that does not mean the hardware is dedicated to your sole use (you can choose to have completely dedicated resources, but that is much more expensive). The public cloud provider shares those resources with other users. Although when you view your VPC, you only see the resources you are attached to, many other users are accessing, for example, the same servers that you are. This is referred to as "multitenancy." The VPC is a snapshot of what resources you are using without showing anyone else's. Without multitenancy, the cloud would not be economically feasible. Multitenancy does present security concerns, but the internal operation of the cloud provider's infrastructure prevents the sharing of information.

The VPC network, although smaller in size, operates exactly like a much larger WAN network. The basic protocols, Ethernet and IPv4, operate exactly as they do in the WAN. Although IPv6 will eventually completely replace IPv4, most public cloud providers have just recently enabled IPv6 in their environment. However, you will probably choose to use IPv4.

Each VM on the server mimics an actual hardware device or software application. For instance, a compute VM, just like a hardware-based computer, has virtual Ethernet interfaces with MAC addresses. IPv4 will connect all the VMs using IPv4 addresses and is responsible for routing between subnets.

Generally, you should enable both public and private subnets. Functions, such as a web server on a compute VM, which need to be accessed from the Internet or other network will reside in a public subnet and need public IP addresses. Resources that should not be accessible from outside of the organization should reside in a private subnet and be assigned private IP addresses. If a user in the organization in the

M. S. Kingsley, *Cloud Technologies and Services*, Textbooks in Telecommunication Engineering, https://doi.org/10.1007/978-3-031-33669-0_4

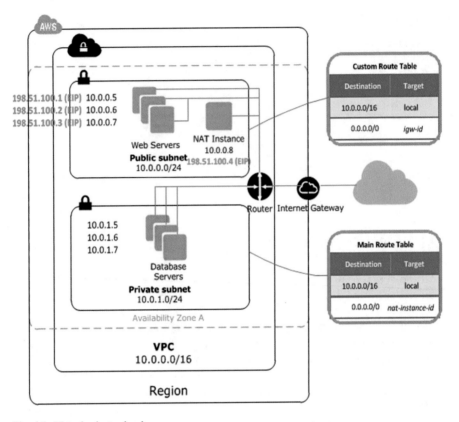

Fig. 4.1 Virtual private cloud

private subnet are allowed to access the Internet, they can do so by accessing to a Network Address Translation (NAT) VM where they are connected to a public IP address in the public subnet to access the Internet.

What may confuse someone with networking experience is that when looking at your VPC via the cloud provider portal, you will not see network components such as switches and routers. The cloud provider has hidden them from your view and takes care of all the internal networking mechanisms after you define the VPC configuration.

Figure 4.1 illustrates a typical VPC environment. Although the example presented is for AWS, it will be similar for any public cloud provider. The VPC implements all of the networking concepts previously discussed. The following describes in detail what is included in the VPC:

1. The VPC you create will be in an AWS region and availability zones that you define. This VPC is implemented in only one availability zone. The Class A private address space of 10.0.0.0/16 is allocated for the entire VPC.

2. The 10.0.0.0/16 address space is further subnetted as /24, which results in the third octet being used for subnets from 10.0.0.0/24 to 10.0.255.0/24. Therefore, there are now 256 subnets available. The fourth octet allows for 256 host addresses for each of the 256 subnets.

3. Only two subnets are defined here: 10.0.0.0/24 for a public subnet and 10.0.1.0/24 for a private subnet.

4. The public subnet 10.0.0.0/24 contains web server VMs which have private IP addresses for use in the VPC 10.0.0.5, 10.0.0.6, and 10.0.0.7). They also have public IP addresses, so the web servers can be accessed from the Internet (198.51.100.1, 198.51.100.2, and 192.51.100.3). These IP addresses are called "Elastic IP Addresses (EIPs)," which is the name that AWS uses for static IP addresses. Web servers require static IP addresses, so the web server IP addresses will not change.

5. The public subnet 10.0.0.0/24 also has a NAT instance, which allows users in the private subnet to reach the public Internet.

6. The VPC has an Internet gateway that allows access to and from the Internet.

7. The diagram shows a router for illustration purposes.

8. However, router and switch functionality, while operational, is usually hidden in a VPC diagram.

9. The private subnet of 10.0.0.1/24 is where the database VMs are placed. The databases have been assigned private IP addresses (10.0.1.5, 10.0.1.6, and 10.0.1.7).

10. The main route table is connected to the private subnet. The entry 10.0.0.0/16 allows the private subnet to communicate with the "local" public subnet. 0.0.0.0/0 allows connections from the private subnet to the NAT instance in the public subnet to reach the Internet.

11. A custom route table is attached to the public subnet. 10.0.0.0/16 allows the public subnet to communicate with the "local" private subnet. 0.0.0.0/0 allows access to the Internet Gateway from the public subnet.

4.2 What's Next?

In Sect. 4.1, the basics of a public cloud infrastructure were investigated. Section 4.2 emphasized how the cloud infrastructure depends on networking and the methods necessary to create a virtual private cloud (VPC). Part III focuses first on the services of the top three public cloud provider platforms: AWS, Microsoft Azure, and the Google Cloud Platform. These discussions are followed by Part IV where you will complete comprehensive labs on each provider's platform to create, configure, and deploy a VPC.

Homework Problems and Questions

1.1. Identify and explain the purpose of the following AWS VPC components:

- EC2
- Public subnets
- Private subnets
- Elastic IP addresses
- NAT instance
- Internet gateway
- Main route table
- Custom route table

1.2. Describe the networking functions and their operation in a virtual private cloud.

1.3. Discuss how a virtual private cloud operates in a public cloud provider's environment.

Part III
Public Cloud Providers

Part 1 presented the basics of the public cloud, and *Part 2* provided the necessary networking knowledge to create your own virtual private cloud (VPC) infrastructure.

Part 3 begins by examining the non-technical requirements for designing a cloud infrastructure based on the AWS Well-Architected Framework as the baseline for presenting the cloud services offered by AWS, Azure, and GCP.

The following chapters are included:

- *Chapter 5: Well-Architected Framework*
- *Chapter 6: Amazon Web Services (AWS)*
- *Chapter 7: Microsoft Azure*
- *Chapter 8: Google Cloud Platform*

Comprehensive hands-on labs for each cloud provider follow in Part IV.

Chapter 5
Well-Architected Framework

5.1 Well-Architected Framework

Cloud architecture design is complex. To assist customers in building and operating secure, reliable, efficient, and cost-effective AWS cloud infrastructures, AWS developed the Well-Architected Framework (WAF).

The WAF can be thought of as the technical mission statement of AWS (AWS, 2022). Although AWS developed the Well-Architected Framework, it has also been adopted by Microsoft Azure with little change (Microsoft, 2022). GCP has also integrated it into their culture with some revisions and consolidation of material (Google, 2022). Therefore, it has become a de facto standard for the public cloud industry.

Whether you work for a public cloud provider or for a customer of one, the WAF is integrated into the cloud engineering culture. Be prepared to understand its contents on cloud certification exams as well as employment interviews with industry participants.

The Well-Architected Framework begins by outlining *general design principles*. These will be followed by a discussion of the six pillars of design. Each of the pillars will be further examined to derive their *individual design principles*, which then motivate design *best practices*.

5.2 Well-Architected Framework General Design Principles

The Well-Architected Framework identifies a set of general design principles that encourage successful cloud infrastructure designs. Prospective cloud engineers should be familiar with the WAF, regardless of the provider under consideration.

© The Author(s), under exclusive license to Springer Nature Switzerland AG 2024
M. S. Kingsley, *Cloud Technologies and Services*, Textbooks in
Telecommunication Engineering, https://doi.org/10.1007/978-3-031-33669-0_5

- *Stop guessing your capacity needs.* On-premises data centers must design resources to meet peak traffic demands. Peaks can occur frequently and randomly, and their duration difficult to predict. The result is that a lot of data resources are idle much of the time. This is costly first because the resource capacity has to be purchased and made operational, and then because it is still using electricity and other support resources while idle. Not designing for peak traffic demand will result in not being able to serve customers during peak periods, which will also result in less revenue than the cost of designing for peak usage.

  Public cloud providers accept the responsibility and cost of providing enough resources for any fluctuation in demand. The necessary resources are quickly provisioned and operational to adapt exactly to the need, while the customer only pays for the use of those resources during the time when they are required. When the demand decreases the resources are decommissioned, and billing for them stops. Guessing how many resources to buy and enable is therefore eliminated when using a public cloud provider.

- *Test systems at production scale.* When provisioning resources in a private data center, testing to see if they operate correctly in a production environment is difficult. Generally, enough equipment to test system operation is not available for testing. Therefore, testing is often bypassed and the resources are directly "cut over" to production, or the online environment. It is not uncommon to begin placing new resources online only to realize there are problems and have to back out. This practice delays adding needed resources, often causes outages, and is an expensive use of personnel.

- Public cloud providers eliminate these problems. Since resources can be quickly turned up in the exact configuration needed and tested using historic traffic workloads, the behavior of the system can be verified or problems detected that do not require affecting production workloads. When the system operation is verified to be error-free, it can then be implemented in the production environment. Charges only apply during the time the cloud resources are operational during verification.

- *Automate to make architectural experimentation (and implementation) easier.* The IT and networking industries have long been, and still are, driven by many manual processes. For example, networking and other equipment require using a command-line interface (CLI) for configuration. It is time-consuming and error-prone. At the scale of the cloud, it is completely ineffective. Automation methods must be implemented to keep up with the scale of the cloud.

- Cloud providers have relieved customers of these ineffective methods by creating pools of resources that are standing by when they are needed. Cloud providers operate in an "Infrastructure-as-Code" environment where resources can be defined in software code, tested for accuracy, and then downloaded into multiple hardware systems inside the cloud provider simultaneously and made operational in production or for testing immediately. Provisioning takes place behind the scenes; the customer is never burdened with this detail. Templates of standard configurations can be created that allow rapid and error-free expansion of any number of desired resources when needed. What used to take days, even weeks, before now can occur in minutes.

- *Allow for evolutionary architectures.* Historically, in the IT and networking environments, new technology developments were rather slow to develop. Four or five years (or longer) was not uncommon before major technological changes were developed and ready for implementation. Very few changes were allowed during the technology's lifetime. Today, technology changes must be embraced and systems operational in days or less. Automation and rapid testing capabilities dramatically shrink the timeline for new technology innovations to quickly become reality in the cloud, greatly enhancing the competitive stance of the cloud provider as well as the customer.
- *Drive architectures using data.* Cloud providers provide a realm of services that allow customers to collect and analyze a vast array of cloud operational data. If the customer adapts to using these resources, they can make fact-based decisions about improving the behavior and operation of the public cloud infrastructure.
- *Improve operations through "game days."* Game days are planned exercises where the architecture, reflecting production scenarios, is tested offline, where problems and inefficiencies can be discovered. Workloads can be stressed, problems inserted to observe their effect, and new ideas tried. As a result, game days can build a knowledge base of solutions and best practices that can be implemented before problems actually occur.

The general design principles are followed by focusing on the Six Pillars of the Well-Architected Framework.

5.3 Six Pillars

Each of the six pillars provides design principles, foundational questions, and best practices used to measure the effectiveness of a cloud architecture. The six pillars are:

1. *Operational excellence*
2. *Security*
3. *Reliability*
4. *Performance efficiency*
5. *Cost optimization*
6. *Sustainability*

Each is investigated further below.

5.3.1 Operational Excellence

Operational excellence is the ability to run systems and monitor operations to achieve business value. There are five design principles for operational excellence in the cloud:

1. *Perform operations as code.* This is the concept of Infrastructure-as-Code (IaC). Instead of configuring and provisioning resources manually, the instructions are written in software. When initiated, the software makes all the necessary functions and connections automatically. This greatly reduces configuration and provisioning time, eliminates human error, and enables consistent operation.
2. *Make frequent, small, reversible changes.* Historically, major IT and network changes rarely occurred. When they did, they often involved "forklift upgrades," meaning the entire system may need to be replaced. As previously discussed, changes today must occur more often. However, changes should be small and easily reversible, avoiding the difficulties of attempting major changes and then having to back out of them, often by inserting problems into the previous version.
3. *Refine operations procedures frequently.* Document procedures as well as problems that have occurred to avoid repeating previous mistakes and to highlight where improvements can be made. Implement ways to consolidate and disseminate learned information.
4. *Anticipate failure.* Analyze infrastructure operations and brainstorm ways in which the systems may fail before they do. Test theoretical failures to see how they respond and how they can be avoided. Document what was learned for future access.
5. *Learn from all operational failures.* Avoid "firefighting." Take time to review the lessons learned from failures. Share what is learned across the organization.
   Further, there are four areas of Operation Excellence best practices:

   1. *Organization.* Understanding your internal and external environment results in designs that support your stakeholders.
   2. *Prepare.* Knowing the requirements of your workloads results in designs that support your work.
   3. *Operate.* Metrics are defined, and operational outcomes are measured.
   4. *Evolve.* Making small and incremental changes creates an environment of consistent improvement.

5.3.2 Security

The security pillar protects data, systems, and company assets. There are seven design principles for well-architected framework security:

1. *Implement a strong identity foundation.* Abide by the principle of least privilege. Use centralized identity management to authorize access to AWS resources.

2. *Enable traceability.* Monitor your environment in real time and send alerts when metrics are exceeded. Send alerts when variances occur, log the detected problems, and take automated actions to correct the issues. Perform scheduled audits.
3. *Apply security at all layers.* Apply defense-in-depth to all layers of the system (VPC, EC2, ELB, operating system, application, code, etc.).
4. *Automate security best practices.* Create secure components with controls that are defined in code and stored in a version control system.
5. *Protect data in transit and at rest.* Classify data by level of sensitivity and apply encryption where appropriate.
6. *Keep people away from data.* Eliminate as much as possible direct access to data to reduce human errors and prevent malicious intent or theft.
7. *Prepare for security events.* Prepare for security events by having an incident management and response plan. Practice incident response to improve incident detection, investigation, and recovery.

There are six best practice areas for security in the cloud:

1. *Security.* This step advocates applying comprehensive security measures to all infrastructure areas.
2. *Identity and Access Management.* IAM ensures only authorized users have access to your resources as defined by you.
3. *Detection.* These controls identify potential security threats that should be investigated.
4. *Infrastructure Protection.* This guarantees systems and services are safe from unauthorized access.
5. *Data Protection.* This includes date classification and recovery, encryption of data at rest and in transit, and other measures to protect data from theft, loss, or corruption.
6. *Incident Response.* Being prepared for security incidents should they occur and limiting their impact.

5.3.3 Reliability

The reliability pillar is responsible for anticipating, detecting, and recovering from failures and automatically initiating remediation. Also, the ability to allocate resources

There are five design principles for reliability in the cloud:

1. *Automatically recover from failures.* Key performance indicators (KPIs) that measure business value should be implemented and monitored. When thresholds are crossed, automate notifications of events and the recovery process when possible.
2. *Test recovery procedures.* Simulate workload failure scenarios to discover vulnerabilities, create recovery plans, and validate their effectiveness.

3. *Scale horizontally to increase aggregate workload availability.* Larger resources are more vulnerable to failures, and their impact is much greater than that of small resources. Replacing large resources with several or many smaller resources limits failure impact and eliminates single points of failure.
4. *Stop guessing capacity.* In the cloud, resources can be quickly and automatically adapted to workloads. Proper application and monitoring of services enables just-in-time management of resources to meet demand or avoid unnecessary spending.
5. *Manage change in automation.* Changes to the infrastructure should be accomplished using automation rather than manual human intervention.

There are four best practice areas for reliability in the cloud:

1. *Foundations.* Foundational requirements are those that span beyond a single workload. An example is having enough available bandwidth to service all workloads.
2. *Workload Architecture.* Includes design decisions for both software and infrastructure. The tools chosen by developers will affect the reliability of workloads across all of the six pillars.
3. *Change Management.* Changes to workloads can occur automatically. If so, they should be monitored and audited. If manually implemented, IAM should be used for access and changes, and their initiator should be logged.
4. *Failure Management.* Systems and service failures will occur. Methods should be in place to detect and as much as possible automatically correct failures. Disaster recovery plans should be in place and tested.

5.3.4 Performance Efficiency

The performance efficiency pillar involves efficiently using computing resources to meet needs and to adjust to demand changes. There are five design principles for performance efficiency in the cloud:

1. *Democratize advanced technologies.* It is difficult to be proficient in the use of over 200 services. Therefore, instead of mastering a technology where your in-house expertise is lacking, consume the technology as a service. Delegate technical details to the cloud provider, who has specialized expertise. This allows you to focus on product development rather than resource provisioning.
2. *Go global in minutes.* Take advantage of AWS's ability to deploy your infrastructure across multiple regions. If done properly, application latency will be decreased and costs minimized.
3. *Use serverless architectures.* "Serverless" does not mean that underneath the cloud hood there are no servers because there are. Serverless means that you as the customer do not have to configure, provision, manage, and maintain physical servers. In addition, serverless applications only create and use resources when

they are required. Therefore, serverless removes the management responsibility of physical servers from the customer as well as reduces costs due to the as-needed usage model.

4. *Experiment more often.* Since resources can be replicated, experimenting with new configurations can be accomplished at a minimal cost and provide extensive insight into infrastructure operation.
5. *Consider mechanical sympathy.* Understand how cloud services are consumed. For example, choose databases based on the structure of your data and match that to the most appropriate database format.

There are four best practice areas for performance efficiency in the cloud:

1. *Selection.* Selection of the best solutions for your environment should be based on the types of workloads applied. Individualized service solutions should be examined rather than just generic ones.
2. *Review.* Scheduled reviews of architecture and create opportunities for improvement through new services or by correcting inefficiencies in existing ones.
3. *Monitoring.* Monitoring is vital so problems can be resolved quickly.
4. *Tradeoffs.* Design choices always have tradeoffs. For example, reliability affects performance. A common tradeoff must often be made between acceptable reliability and the cost required to deliver it.

5.3.5 Cost Optimization

The cost optimization pillar focuses on running systems to cost-effectively deliver business value. There are five design principles for cost optimization in the cloud:

1. *Implement cloud financial management.* Most cloud customers do not manage their infrastructure costs efficiently. Much of this waste is due to unused or underused services not taking advantage of cost-saving plans available. Cloud providers offer many cost optimization tools that can dramatically decrease costs. It is imperative that cloud customers learn and apply these tools to their operations.
2. *Adopt a consumption model.* Pay only for the resources you need at any given moment and increase or decrease resources based on real-time workload demands.
3. *Measure overall efficiency.* Track overall efficiency by tracking the usage and costs of services provided to users of those resources. Identify and correct wasteful resource usage.
4. *Stop spending money on undifferentiated heavy lifting.* Allow the cloud provider to assist in areas where the customer has less expertise.
5. *Analyze and attribute expenditure.* Accurately track the usage and costs of systems and then chargeback costs to the users of consumed services. This encourages the service users to more efficiently use the services they need.

There are five best practice areas for cost optimization in the cloud:

1. *Practice cloud financial management.* Optimize resource cost and usage.
2. *Expenditure and usage awareness.* Knowing where to monitor and allocate costs can result in overall cost conservation.
3. *Cost-effective resources.* Right-sizing services, reserve instances, and savings plans are examples of cost effective selections of resources.
4. *Manage demand and supply resources.* Take advantage of the pay-as-you-go cost basis and use autoscaling to adjust to demand variances.
5. *Optimize over time.* Using continuous monitoring allows analysis of costs, and adjustments can be made to optimize costs over time.

5.3.6 Sustainability

Sustainability is a recent addition to the AWS pillars and may not be reflected in some AWS as well as Microsoft Azure and Google Cloud Platform documentation.

Although focused on energy use and emission reduction, the sustainability pillar is concerned with efficiency at every level of the enterprise.

There are six design principles for the sustainability of cloud workloads:

1. *Understand your impact.* Measure and analyze the impact of your cloud workloads, from provisioning to decommissioning.
2. *Establish sustainability goals.* Establish sustainability goals for each workload. Model the return on investment (ROI) of each goal.
3. *Maximize utilization.* Design workload to maximize use and energy efficiency.
4. *Anticipate and adopt more efficient hardware and software offerings.* Motivate suppliers and vendors to develop energy- and cost-efficient products.
5. *Use managed services.* Managed services shift much of the management and operation of services from the customer to the cloud provider who has expertise in creating efficient services.
6. *Reduce the downstream impact of your cloud workloads.* Reduce the energy or other resources used by your services. Reduce or eliminate user requirements to upgrade or maintain your services.

The following steps should be used for continuous sustainability improvement:

1. *Identify targets for improvement.* Review workloads to identify areas for sustainability improvement.
2. *Evaluate specific improvements.* Evaluate targeted areas for improvement against cost and risk factors.
3. *Prioritize and plan improvements.* Choose areas where the most improvement can be accomplished and focus on their accomplishments first.
4. *Test and validate improvements.* Test improvements against criteria for evidence that goals have been accomplished.

5. *Deploy changes to production.* Once the proposed changes have been confirmed, then apply to the production environment.
6. *Measure results and replicate successes.* Monitor changes to verify positive results. Use this process to iteratively continue improvement in other areas.

5.4 Summary

The AWS Well-Architected Framework (WAF) provides guidelines for the design of cloud-based architectures. It provides general cloud design principles and then further details design principles and best practices for six pillars, or categories, under the WAF.

Developed by AWS, the WAF has been adopted by Microsoft Azure and the Google Cloud Platform with few revisions and has therefore become a de facto standard for the industry. New industry participants should be familiar with the six pillars when attempting certification exams as well as when preparing for an interview with a public cloud provider since they are an integral part of their engineering culture.

Homework Problems and Questions

1.1 Describe the AWS Well-Architected Framework's general design principles.
1.2 List and define the functions of the six WAF pillars.
1.3 Choose one of the six WAF pillars and discuss in detail its design principles and best practices.
1.4 Discuss how load balancing and auto-scaling relate to the six pillars.

Bibliography

AWS. (2022). *AWS Well-Architected Framework* [Online]. Available at: https://docs.aws.amazon.com/wellarchitected/latest/framework/welcome.html

Google. (2022). *Google Cloud Architecture Framework* [Online]. Available at: https://cloud.google.com/architecture/framework

Microsoft. (2022). *Microsoft Azure Well-Architected Framework* [Online]. Available at: https://learn.microsoft.com/en-us/azure/architecture/framework/

Chapter 6
Amazon Web Services (AWS)

6.1 AWS Global Infrastructure

AWS is globally connected, serving 245 countries (Anon., 2021). The AWS infrastructure is a flexible, reliable, scalable, and secure cloud computing environment. The following are features of the AWS infrastructure:

- **Elasticity.**
- **Scalability.**
- **Fault-tolerance.**
- **High availability.**

The AWS infrastructure is supported by regions, availability zones (AZs), edge locations, and local zones.

6.1.1 AWS Regions

The top of the AWS hierarchy is the region. Regions contain clusters of data centers. Each region is independent and isolated from the others. As of this writing, there are 25 AWS regions with more planned. Regions are identified by a location as well as a region code. Below is a partial list of AWS regions as examples (Fig. 6.1):

Not all regions offer every AWS service. Further, pricing for a service may differ from region to region. It is desirable to use services that reside in a region closest to the customer's physical location to minimize latency. However, it may be more economical to attach services that reside in other regions.

M. S. Kingsley, *Cloud Technologies and Services*, Textbooks in
Telecommunication Engineering, https://doi.org/10.1007/978-3-031-33669-0_6

Fig. 6.1 AWS regions

Name	Code Name
US East (N. Virginia)	us-east-1
US East (Ohio)	us-east-2
US West (N. California)	us-west-1
US West (Oregon)	us-west-2
Canada (Central)	ca-central-1
EU (Ireland)	eu-west-1
EU (Frankfurt)	eu-central-1
EU (London)	eu-west-2
Asia Pacific (Tokyo)	ap-northeast-1
Asia Pacific (Seoul)	ap-northeast-2
Asia Pacific (Singapore)	ap-southeast-1
Asia Pacific (Sydney)	ap-southeast-2
Asia Pacific (Mumbai)	ap-south-1
South America (São Paulo)	sa-east-1

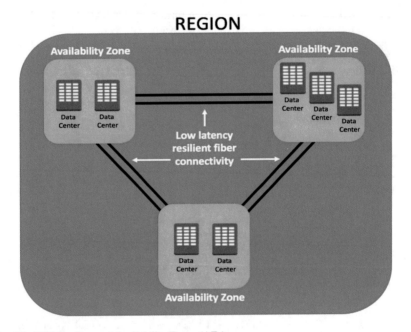

Fig. 6.2 AWS Regions and Availability Zones (AZs).

6.1.2   AWS Availability Zones

Each region contains between two and five *availability zones (AZs)*, each of which contains one or more physical data centers where AWS hardware that hosts services physically resides (Fig. 6.2).

An AWS data center typically houses up to 80,000 servers. Each AZ, as well as the individual data centers, are physically isolated from the others with redundant power and network connectivity. Many AWS services are replicated, either automatically or by customer choice, in one or more AZs. If an AZ fails, there is at least one other AZ where the services will still be operational.

Availability zones are connected by diverse, high-speed fiber optic network connections that provide high availability and low latency.

Currently, there are 81 AWS availability zones.

Availability zones are identified by a region code with an additional AZ designator. For example, the AZs within the eu-west-1 region (EU Ireland) are eu-west-1a, eu-west-1b, and eu-west-1c.

6.1.3   AWS Edge Locations

AWS has hundreds of edge locations located in major cities all over the world to provide rapid data delivery to customers. (Scott, AWS) However, services do not currently originate from edge locations. Instead, commonly accessed information is cached, or stored, at the edge locations. For instance, Netflix delivers movies using AWS. Without the edge location, downloading a movie would have to be retrieved from a more distant data center. Instead, the most popular movies are cached at edge locations and delivered via the AWS *Content Delivery Network (CDN)*, where they can be more rapidly downloaded by customers.

6.1.4   AWS Local Zones

It is now possible for hardware servers that run compute, storage, and database services close to or on customer premises to deliver sub-millisecond latency response.

Local zones are a new service that is currently only available in a few metropolitan areas, but many more are planned.

6.1.5   AWS Global Network

AWS is supported by a global fiber optic network operating at 100 billion bits per second that is redundant and provides low-latency data transfers (Fig. 6.3). Counties are connected by trans-oceanic cables across the Atlantic, Pacific, and Indian oceans, as well as the Mediterranean, Red, and China seas.

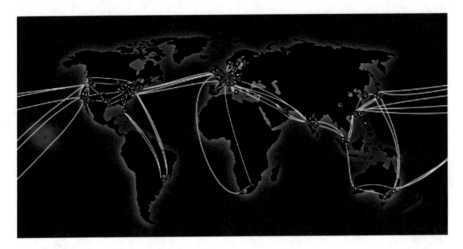

Fig. 6.3 AWS Global Network. (Souk, 2022)

6.2   AWS Services

AWS offers a comprehensive menu of services, and more are continually being developed. Over 200 services are available in the following categories (Anon., 2023):

- Networking and content delivery.
- Computing.
- Storage.
- Databases.
- Analytics.
- Security, Identity, and Compliance.
- Management tools.
- Application services.
- Developer tools.
- Mobile services.
- AR and VR.
- AWS cost management.
- Blockchain.
- Enterprise apps.
- Customer engagement.
- End-user computing.
- Game Technology.
- Internet of Things.
- Machine Learning.
- Migration and transfer.
- Robotics.
- Satellite.

It would be very difficult to master all the services available on AWS. In fact, many engineers specialize in one or a few services. We will look at the following basic service categories in detail:

- Management and Monitoring.
- Security, Identity, and Compliance.
- Networking.
- Autoscaling.
- Elastic Load Balancing.
- Compute.
- Storage.
- Database.

6.2.1 Management and Monitoring

Many AWS services are supported by the management and monitoring of your AWS infrastructure.

6.2.1.1 Management Console

The AWS Management Console is a web-based portal to access your AWS cloud infrastructure. It is the command center for all of your AWS management tasks, including configuration of services and observing infrastructure operation.

AWS Management
Console

6.2.1.2 CloudWatch

Cloudwatch is an AWS monitoring service. It collects performance statistics based on metrics for AWS resources. Alarms are generated when metrics are not met.

Amazon CloudWatch

An example of Cloudwatch's usefulness is that it monitors web servers. When traffic thresholds are reached, it informs autoscaling, which then adds or deletes web servers.

CloudWatch has a basic set of features available at no charge. More advanced features can be accessed for additional cost.

6.2.2   Security, Identity, and Compliance

Many AWS services support the security, identity, and compliance of a customer's AWS infrastructure. The most important is Identity and Access Management (IAM).

6.2.2.1   Identity and Access Management

AWS has a comprehensive offering of security services, including encryption, to secure your services and resources. Other security capabilities are built individually into various services as well. However, the basic AWS security service is **Identity and Access Management (IAM)**. It is the starting point for building AWS's functional infrastructure.

AWS Identity &
Access Management

IAM provides **authentication** and **authorization** to users or services. Authentication is the process of verifying who you are and that you can access AWS. The most basic authentication methods are usernames and passwords. In addition to a username and password pair, IAM has other tools such as *Multi-factor Authentication (MFA)*, wherein a secondary authentication method is required. Examples of MFA include a callback to a stored telephone number, biometric scanning of a fingerprint, or the use of third-party authentication tools such as the Google Authenticator ©.

Authorization determines what you can access within the AWS environment and is accomplished using *policies.* A policy is a list of what a user or resource can access and what they can do when they do access it.

IAM applies security via users, groups, and roles to manage permissions to allow or deny access to AWS resources. When first creating an AWS account, the *root* user is created. The root user can do anything within AWS. Obviously, this should not be allowed. Therefore, the root user should create other non-root users. In fact, even the administrator should create a user account for themselves since using the powerful root user privileges can result in unintended consequences.

Users are generally people. Each person is provided access based on an access policy. For example, a salesperson may only be allowed read access to data in a database, whereas the sales manager probably requires write access as well to change product prices.

AWS has a large library whereby prewritten policies can be selected and applied, or the administrator can create customized policies if desired.

Creating policies for individual users can be a difficult and time-consuming task. Therefore, users can be assigned to *groups* that apply a specific policy to a group of similar users such as salespeople or development engineers.

Roles are similar to those of IAM users, but there are distinct differences. IAM users are used to allow external access to the AWS environment. On the other hand, roles are intended primarily to allow access inside AWS. Also, whereas IAM users are usually people, roles are created to allow or deny access to resources from another resource. For example, a compute instance may require access to a database VM. This access must be secured as well. Therefore, the compute instance will be assigned a role that allows access to the database instance. A role can be assigned to an IAM user to allow the user temporary access to AWS resources. A user could be allowed access to a storage instance for a short duration of time, say a few minutes, and the permission would then be revoked.

6.2.3   AWS Data Center Networking

Inside AWS data centers are massive fiber optic networks that connect data center servers and other hardware. However, software-defined, or "virtual," networks operate inside hardware servers.

6.2.3.1   Virtual Private Cloud (VPC)

The heart of your AWS infrastructure is the **Virtual Private Cloud (VPC)**. It is your small slice of the AWS infrastructure and where many of your compute, storage, database, and other services reside.

Virtual Private Cloud (VPC)

VPC is also the networking heartbeat that connects services inside and outside of your VPC. The VPC network is virtual but operates just like a physical network. For example, VMs in a VPC have Ethernet and IP addresses just like physical networks. For small infrastructures, in-depth networking knowledge is not necessary. AWS will make all your network assignments necessary if requested. However, much more care must be taken to properly configure a larger VPC network. Unlike a physical network, IP addresses cannot be changed once they have been assigned to a resource. Sloppy IP address and route planning can lead to stranded resources and traffic bottlenecks that require deleting and reconfiguring the VPC network. Resulting in downtime, irate customers, and lost revenue. It is here that a cloud engineer must have more extensive networking knowledge.

The main components of a VPC are:

- *Subnets* separate public-accessible resources like web servers from private resources like storage systems and databases.
- An *Internet gateway* provides access to and from the Internet.
- Administrator-configured *route tables* determine traffic routing in the VPC and to and from the Internet.
- *Network Address Translation (NAT)* allows access to the Internet for private resources such as databases that need to download updates.
- Compute, storage, database, and other functions as needed.

6.2.3.2   Elastic Load Balancer (ELB)

Elastic Load Balancing is used to evenly balance traffic across AWS services. For example, multiple EC2s are used to host web servers for a busy website. Incoming web traffic may favor one web server until it is fully utilized, resulting in a slower response time without load balancing. ELB evenly distributes the web requests across all available EC2 web servers, greatly improving performance and decreasing latency.

There are three types of AWS load balancers.

- *Application Load Balancer (ALB)* operates at Layer 7 of the OSI model and distributes traffic based on HTTP and HTTPS as well as other content information.
- *Network Load Balancer (NLB)* works at Layer 4 of the OSI model and directs traffic based on TCP and UDP. The NLB can handle millions of requests per second with very low latency.
- *Classic Load Balancer (CLB)* is an early AWS offering that operates at Layer 4 through Layer 7 of the OSI model. It is only used with legacy EC2 instances.

Elastic Load Balancing

6.2.3.3 Autoscaling

Traditionally, data center resource planning is inexact. If administrators underestimate network bandwidth and data center resources, then requests would be rejected in times of high demand, resulting in lost revenue and customer dissatisfaction. Therefore, administrators often opt to *overprovision* or design resources to handle the highest demand peaks. However, building resources, for example, to service an annual clearance sale would meet the need for that short period of time but would result in wasted and expensive capacity for the rest of the year.

AWS autoscaling enables real-time dynamic resource allocation based on demand. For example, as web traffic increases, AWS will add web servers; as it decreases, web servers will be removed. With autoscaling, resource planning is no longer a guessing game. Demand and capacity are perfectly matched. The bonus is that you only pay for AWS resources that you actually use.

AWS Auto Scaling

6.2.4 Compute

AWS data centers have tens of thousands of servers. However, customers need only define and use how much computing and related resources, such as memory and storage, they need now. If more compute capacity is needed, it can be added quickly or disabled if the need decreases. AWS allocates compute resources using Elastic Cloud Compute (EC2).

6.2.4.1 Elastic Cloud Compute (EC2)

Usually the first service required in your AWS infrastructure is compute capability. Although AWS has many services in the compute category, the most used is *Elastic Cloud Compute (EC2)*. EC2 instances are placed inside your VPC. EC2 is a virtual machine (VM), or software computer, that resides on a physical server.

In large website implementations, a physical server can host dozens, even hundreds, of EC2 instances.

Amazon EC2

Although we focus on the three-tier web cloud-based architecture with the web server as an application running on EC2, literally hundreds of applications can be similarly hosted on EC2.

EC2 is an "unmanaged" AWS service. The administrator must make many choices regarding the capabilities of an EC2 based on its intended use as well as configure those capabilities. Fortunately, those configurations can be automated using a configuration template called an Amazon Machine Image (AMI).

EC2 AMIs

Just like your laptop or workstation computer, EC2 requires processors, an operating system, memory, storage, and networking to connect to other AWS services. Before turning on an EC2, all these components must be defined.

AWS provides a library of hundreds of preconfigured combinations of processors, operating systems, memory, and storage capabilities. These preconfigured components, or templates, are called *Amazon Machine Images (AMIs)*. The "image" in this case is not like a photograph, but a template of components that create the capabilities of an EC2. You can select small AMIs with a few processors and a little memory and storage or very large AMIs with dozens of processors and huge memory and storage capability. You pay proportionally more or less depending on the AMI you select. You can access the AWS AMI library, choose AMIs created by others in the AWS Marketplace, or create your own custom AMI. The same AMI can be used to launch any number of identical EC2 instances.

Each AWS class has AMIs of various compute, storage, operating system, and memory configurations of all sizes that can be selected to meet the various EC2 requirements.

AWS offers over 400 different AMIs. The following describes the five EC2 AMI classes:

- *General purpose.*

AWS general purpose virtual machines provide a balanced CPU-to-memory ratio. They are ideal for testing and development, small to medium databases, and low to medium traffic web servers.

- **Compute optimized.**

AWS compute optimized virtual machines are designed with a high CPU-to-memory ratio. They can be used for medium traffic web and batch processing.

- **Memory optimized.**

AWS memory optimized virtual machines are designed with a high memory-to-CPU ratio. The work well with relational databases and in-memory caches.

- **Storage optimized.**

AWS storage optimized virtual machines provide high disk throughput and are ideal for transactional workloads and data warehouses.

- **Accelerated computing.**

AWS accelerated computing is used for Artificial Intelligence (AI), Machine Learning (ML), and High-Performance Computing (HPC).

EC2 Reliability and Scalability

EC2 is reliable and scalable. If an EC2 fails there is a backup that will immediately take over the workload. If many EC2s are used for the same function such as a web site, AWS can load balance the traffic across all EC2s. Finally, as in the case of a web server that is serviced by numerous EC2s, as the traffic to the web site increases or decreased AWS will automatically increase the number of EC2s to accommodate more traffic of or decrease EC2s to save money when traffic lessens.

Amazon CloudWatch

EC2 On-Demand, Reserved, and Spot Instances

There are several ways EC2s can be delivered and billed. *On-demand* instances are continuously available and require no long-term commitment. However, they are also the most expensive. There are more cost-effective solutions.

Knowing your workload requirements is important before moving your infrastructure to AWS. Guessing resources and network bandwidth is unwise and expensive. Over estimating will be unnecessarily expensive while underestimating will result in poor service, angry customers, and lost revenue. There are many tools that can monitor workload demand and traffic which should be used to size your EC2s.

If resource workloads are known and it is anticipated they will remain at that level for the long term, it is more cost-effective to use *reserved* instances. With reserved instances you contract with AWS for a one or three years results in up to 70% cost savings over on-demand instances.

In general, you should target 90% of your EC2 resources as reserved instances . On-demand instances should only be used for testing applications or for unscheduled or out of sequence jobs and target the remaining 20% of the number of EC2 instances.

Another EC2 pricing option is to use *spot* instances. Spot instances are requested and made available to you when there is unused resources available from AWS. However, spot instances can be turned off if AWS needs the spot resources you have been assigned. If AWS intends to shut off your spot instance a spot is instance they will only give you 2 min notice.

Spot instances can be used for jobs that are not time critical, that can be interrupted and restarted without problems, or that cost savings is critical. Spot instances are up to 90% cheaper than on-demand instances (Anon., n.d.-a, n.d.-b).

Other Parameters

The details of several parameters are expanded below:

- *Operating system options:* Operating system choices available for AWS AMIs include various versions of Linux and Windows. In addition, AWS is the only public cloud provider that supports macOS.
- *Processor options:* AWS AMIs supports Intel, AMD, and ARM-based processors.
- *EC2 limits:* Amazon EC2 VMs scale up to 448 vCPUs and 24,576 GB of RAM.

6.2.5   Storage

Just like your workstation or laptop, EC2 virtual computers require storage. Storage can be either on a Hard Disk Drive (HDD) or a Solid-State Device (SDD) devices. When enabling a service, you will need to define what storage will be used. AWS offers instance store, file, block, and object storage options.

6.2.5.1   EC2 Instance Store

All EC2 instances have some on board storage capability by default. How much depends on the size of the instance.

Instance store storage in is "ephemeral" which means "temporary." In computer lingo it means it is "non-persistent," or not permanent. If the EC2 is stopped or fails the data in instance store storage is deleted and irretrievable. You will want to add other storage solutions to keep important data as long as desired.

So why would you even use ephemeral storage? It is useful to store short duration data. Example are shopping cart information or temporary backups.

6.2.5.2   Elastic File Service (EFS)

EFS is a file-based system that uses NFS version 4. It is similar to the file system that is used on a personal computer with folders and underlying files. EFS can be accessed simultaneously by multiple EC2s providing a common file system for multiple applications running on different EC2s.

Amazon Elastic
File System

EFS offers consistently low latency and redundancy. EFS data is replicated across multiple availability zones. If a failure occurs, EFS has multiple copies of your data that can be accessed.

EFS is fully managed by AWS so you have limited maintenance responsibility.

6.2.5.3   Elastic Block Storage (EBS)

EBS is persistent block storage for use with EC2. Unlike instance store storage, persistent data will survive if the EC2 is shut down.

EBS is well suited for applications that require access to raw, unformatted, block-level storage. EBS volumes are created in an Availability Zone and then attached to a specific EC2 in your VPC. EBS is usually attached to a single EC2. However, added features allow up to 16 EC2s to access EBS.

Amazon Elastic Block Store

EBS data is automatically replicated in the Availability Zone in which it resides.

There are four categories of EBS based on achieving maximum Input-Output Operations per Second (IOPS) in Mbps:

- *General Purpose SSD* for acceptable IOPS performance.
- *Provisioned IOPS SSD* for I/O intensive workloads.
- *Throughput Optimized HDD* is low cost magnetic storage used for throughput intensive applications.
- *Cold HDD* low cost magnetic tape storage for infrequently accessed data.

6.2.6   Simple Storage Service (S3)

S3 is the AWS persistent object storage solution. Any data format can be stored in S3 including files, photographs, or documents.

Data stored in S3 is automatically replicated in at least three different Availability Zones which guarantees your data will always be available even if two Availability Zones have failed. A disaster of that magnitude is (hopefully) unlikely.

Amazon Simple
Storage Service (S3)

S3 data is stored in a "bucket."
There are five classes of S3 service:

- *S3 Standard* general-purpose storage for frequently accessed data. It is the most expensive S3 storage class.
- *S3 Standard Infrequent Access (IA)* for long-lived but infrequently accessed data. Less expensive than S3 Standard.
- *S3 One Zone Infrequent Access (IA)* where less critical or temporary data is stored in only one availability zone. It is less expensive than S3 Standard IA.
- *S3 Glacier* for archiving data. Very low cost S3 storage but charges apply to retrieve data. Retrieval time minimum is several hours.
- *S3 Glacier Deep Archive* is the least expensive S3 service. However, retrieval time minimum is 12 h.

In general, the quicker you need your data the more your storage will cost per gigabyte but the less it will cost to retrieve it. Long term shortage such as Glacier is economical to store data but much more expensive to retrieve it.

6.2.6.1 S3 Object Lifecyle Management

Data in S3 can be moved from on class to another as defined by the administrator. For instance, after data in S3 Standard can be automatically moved to S3 Standard IA after a specified time such as 30 days; after 60 days there the data can be moved S3 Glacier; after 180 days in Glacier it can be then deleted in desired. Also, *S3 Intelligent Tiering* is available to automatically move data into an out of classes as necessary to maximize storage savings.

6.2.6.2 S3 Residency

Unlike EBS and EFS, S3 does not reside in your VPC but in an AWS Availability Zone. S3 data can be retrieved from anywhere including the Internet.

6.2.6.3 S3 Capacity

S3 can support millions of data requests simultaneously. It will also scale to unlimited capacity immediately and automatically as needed.

6.2.6.4 S3 Redundancy

S3 is extremely redundant. S3 saves a copy of your data in each of at least three different AZ except for S3 Standard IA which provides redundancy in a single AZ. If a failure occurs, S3 has two other copies of you data that can be accessed.

6.2.7 Databases

AWS offers a variety of databases but the most used are relational and non-relational databases. A relational database is similar to a spreadsheet where the pages, or tables, are connected.

Relational databases are very structured. Making changes requires careful planning. They are best used for traditional Customer Resource Management (CRM) or Enterprise Resource Planning (ERP) system where they are queried using the Structured Query Language (SQL) which is a robust computer language designed for managing data in a relational database. Changes to the data structure never or hardly ever occur.

In contrast, non-relational databases, or the NoSQL (pronounced "No SQL") database format is unstructured and ideal for storing documents. Being unstructured, the rows and columns in the tables can be changed on-the-fly. They are very fast and can handle literally millions of requests per second. They are ideal for busy web sites or mobile applications.

6.2.7.1   Relational Databases

AWS's primary relational database service is Relational Database Service (RDS).

Relational Database Service (RDS)

RDS is an AWS relational database service. It is a fully managed service after initial configuration so all maintenance and upgrades are accomplished by AWS.

You can think of RDS as a box for a conventional database. Database engines that can run on RDS are:

- MySQL.
- MariaDB.
- PostgreSQL.
- Oracle.
- Microsoft SQL Server.
- Amazon Aurora.

MySQL, MariaDB, and PostgreSQL are open source databases. Oracle and Microsoft Server can be expensive. The cost of these database is included in the price of RDS.

RDS user configuration includes choosing a database engine and the instance size, security, storage and other parameters.

ECWRDS can be configured to include a standby redundant database in a separate AZ. If the primary database fails the redundant system in the other AZ will take over.

Amazon RDS

RDS/Aurora

Aurora is a fully managed AWS proprietary database based on RDS. It is five times faster than MySQL and ten times cheaper than Oracle! Recently, Amazon replaced all of their internal databases previously using Oracle with Aurora.

Aurora is also highly redundant. It keeps two data copies in each of three different AZs for a total of six copies of your Aurora database.

6.2.7.2 Non-relational Database

Relational databases are highly structured and do not scale well. Many services require the ability to work with data that is not structured and to complete millions of requests per second. Although AWS offers several non-relational database services each focusing on a specific application, DynamoDB meets most needs.

6.2.7.3 DynamoDB

DynamoDB is an AWS fully managed non-relational database. It provides extremely fast and predictable performance and can scale automatically to quickly store and retrieve massive amounts of data. It can service millions of requests per second without delay or degradation. Response times are in the single-digit milliseconds timeframe which can be decreased to nanoseconds using DynamoDB Accelerator (DAX).

Amazon DynamoDB

DynamoDB global tables replicate data across three AZs and automatically scales capacity to accommodate any workloads.

6.3 AWS Lab

You are now ready to experience AWS using the AWS Free Tier. Free tier offers twelve months of free service for dozens of AWS services.

AWS Free Tier is generous but each AWS service has limits. For example, AWS EC2 can be used free for 750 hours which is more than enough time to practice using EC2. Other services have similar usage limits.

After signing up for an AWS Free Tier account you will then create a basic AWS infrastructure that applies what you have learned up to this point (Fig. 6.4). Your design is a basic fully operational three-tier web application running on the AWS platform using the following services:

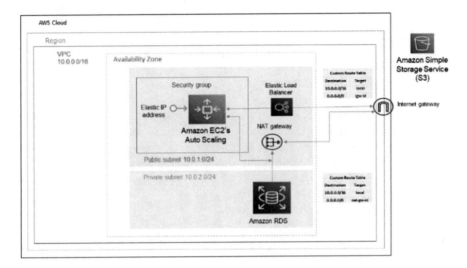

Fig. 6.4 AWS Lab

- AWS Regions and Availability Zones.
- Identity and Access Management.
- VPC network including public and private subnetting.
- EC2 compute.
- S3 storage.
- RDS database.
- Internet gateway.
- NAT gateway.
- Route tables.
- Elastic Load Balancing.
- Autoscaling.
- Security groups.

6.4   Summary

AWS has a global infrastructure that spans regions, availability zones, edge locations, and local zones that support over 200 AWS services with more being added regularly.

Learning all the services offered by AWS would be very difficult so this focus was on the most necessary and used services including:

- Management and Monitoring.
- Security, Identity, and Compliance.
- Networking.
- Compute.

- Storage.
- Database.
- Elastic Load Balancing.
- Autoscaling.

The next step is to integrate these services together into an operational infrastructure in the lab.

Homework Problems and Questions

1.1 Describe the AWS global infrastructure and its related components.

1.2 Describe the following service categories and the individual services they contain:

- Management and Monitoring.
- Security, Identity, and Compliance.
- Networking.
- Compute.
- Storage.
- Database.
- Elastic Load Balancing.
- Autoscaling.

1.3 Differentiate between managed and unmanaged for the following AWS services:

- IAM.
- VPC.
- ELB.
- Autoscaling.
- EC2.
- EFS.
- EBS.
- S3.
- RDS.
- Aurora.
- DynamDB.

1.4 Draw an AWS VPC containing all of the above services.

Bibliography

Anon. (2021). *The deceptively simple origins of AWS*. [Online]. Available at: https://www.aboutamazon.com/news/aws/the-deceptively-simple-origins-of-aws

Anon. (2023). *Amazon*. [Online]. Available at: https://www.aboutamazon.com/what-we-do/amazon-web-services

Anon. (n.d.-a). *Spot by NetApp*. [Online]. Available at: https://spot.io/what-are-ec2-spot-instances/?utm_campaign=Ocean+search&utm_term=amazon%20spot%20

instances&utm_source=adwords&utm_medium=ppc&hsa_ver=3&hsa_kw=amazon%20
spot%20instances&hsa_cam=12797893166&hsa_tgt=kwd-341551341062&hsa_
acc=8916801654&hsa_mt=b

Anon. (n.d.-b). *What aew AWS spot INstances?*. [Online]. Available at: https://spot.io/what-
are-ec2-spot-instances/?utm_campaign=Ocean+search&utm_term=amazon%20spot%20
instances&utm_source=adwords&utm_medium=ppc&hsa_ver=3&hsa_kw=amazon%20
spot%20instances&hsa_cam=12797893166&hsa_tgt=kwd-341551341062&hsa_
acc=8916801654&hsa_mt=b

Souk, A. (2022). *Well-architecting online applications with CloudFront and AWS global accelerator*. [Online]. Available at: https://aws.amazon.com/blogs/networking-and-content-delivery/
well-architecting-online-applications-with-cloudfront-and-aws-global-accelerator/

Chapter 7
Microsoft Azure

7.1 Azure Global Infrastructure

The Microsoft Azure global infrastructure spans over 140 countries and is connected by an expansive, high speed fiber optic network. Components of the Azure cloud include:

- Data centers.
- Geographies.
- Regions.
- Region pairs.
- Availability zones.
- Fiber optic network.

7.1.1 Azure Data Centers

Azure has over 200 data centers globally and more are planned. Each contains thousands of servers and other equipment. Azure data centers are designed for reliability with redundancy for power and other support systems.

Each data center is supported by a massive internal network with thousands of miles of fiber optic cables used to communicate between data center equipment and the external world over the public Internet or private networks.

7.1.2   Geographies

International laws and customer requirements often define data residency and compliance boundaries and limit what countries data can be sent to or stored in. Azure Geographies ensure their customer data is isolated as well as protected from possible privacy and other violations.

7.1.3   Regions

Azure has over 60 regions. A region is a group of data centers. Not all regions offer every Azure service and service pricing for the same service may vary between regions. For instance, the region that is the shortest distance to a customer may not be the least costly.

7.1.4   Region Pairs

Redundancy of Azure regions is accomplished using region pairs. Each region in a geography is paired with another region in the same geography. The pairs are designed such that if one region of the pair goes down the other region will survive and customers will continue to be served. Region pairs are also designed such that upgrades or maintenance in one region will not occur at the same time in both regions.

7.1.5   Availability Zones

Azure Availability Zones are distinct areas within a region that protect against single points of failure. Each region contains at least two availability zones; each availability zone contains at least one data center with independent support systems including power and network connectivity. Availability Zones are designed so that if one Availability Zone fails, all services in that zone will be supported by the other Availability Zone in the region.

7.1.6   Microsoft Network

The Azure global network (Fig. 7.1) consists of 165,000 miles of fiber optic cables, each operating at 100 gigabits per second. Once data enters the Azure network, it never touches the public IP network (Azure, 2023).

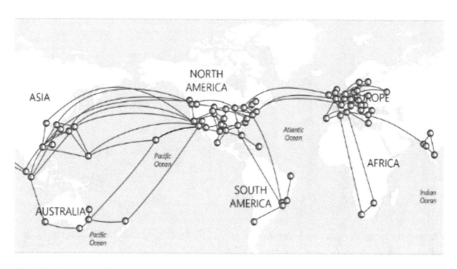

Fig. 7.1 Microsoft fiber optic network

7.2   Azure Services

Like AWS, Microsoft Azure offers a comprehensive menu of over 200 services in the following categories: (Microsft Azure, 2023):

- Compute.
- Networking.
- Storage.
- Internet of Things.
- Mixed reality.
- Integration.
- Identity.
- Security.
- DevOps.
- Migrate.
- Web.
- Mobile.
- Containers.
- Database.
- Analytics.
- Blockchain.
- AI + machine learning.
- Monitor.
- Management + governance.
- Intune.

Only the following core service categories will be examined in detail in the following order:

- Monitor.
- Identity.
- Networking.
- Compute.
- Storage.
- Database.

7.2.1 *Monitor*

Azure services need to be monitored constantly to ensure they are operating and to alert other systems or administrators when problems occur.

7.2.1.1 Azure Monitor

Azure Monitor assists administrators in improving the availability and performance of applications and services. It provides visibility of Azure applications, infrastructure, and networks by collecting, analyzing, and acting upon telemetry data received from customer on-premise and Azure resources. By using machine learning–driven diagnostics and automated actions, it quickly identifies and resolves problems.

Azure Monitor

Azure Monitor is not a free service (Amazon, 2023). You will be charged for how much log data enters the systems as well as for the length of time that data is stored.

7.2.2 *Identity*

Security is essential for cloud services. "Identity" is authenticating that users are who they say they are and authorizing them to access only the appropriate levels of system access necessary for them to accomplish their job.

7.2.2.1 Azure Active Directory (AD)

Azure leverages Microsoft's long-standing Active Directory product to create a multi-tenant, cloud-based identity and access management service. It is a single identity platform that enables single sign-on and multifactor authentication (MFA). It allows the creation of users, groups, and roles. Administrators can limit or allow users access to resources based on many criteria.

Azure Active Directory

7.2.3 Networking

Similar to other public cloud providers, Microsoft Azure focuses on network service virtualization.

7.2.3.1 Azure Virtual Network (VNet)

Azure Virtual Network (VNet) is a customer's small part of the resources in a Microsoft data center. Inside a VNet, the customer will create a variety of virtual machines such as compute, storage, and database.

Azure Vnet

VNets can be created by administrators or automatically by Azure. A VNet can span multiple physical locations to provide redundancy for virtual machines.

VNets provide for isolation between public and private resources as defined by subnets and for automatic routing of traffic both internally and externally across customer's virtual machines in Azure, remote customer resources, and the Internet.

VNet emulates a physical networking infrastructure in the virtual environment under the customer's control. Therefore, a VNet network operates like any other network. VMs have virtual NICs with Ethernet MAC addresses and public subnets allow access to web servers, while private subnets isolate resources from public access. Network Address Translation (NAT) allows resources in private subnets to access the Internet.

VNet uses Azure Active Directory to enforce user access privileges. Further security is accomplished using network security groups that filter inbound and outbound traffic based on administrator requirements.

7.2.3.2   Azure Load Balancing

Load balancing shares the received web traffic between multiple virtual machines. Traffic can be shared equally or unevenly allocated if desired.

Azure load balancing is applied by Azure in two ways: global or regional, and HTTP(S) or non-HTTP(S).

Global load balancing distributes traffic across regional "backends," which are web application deployments in a region, to the closest available backend. Regional load balancing is accomplished across virtual networks or virtual machines.

HTTP and HTTP(S) load balancing operates at the OSI Layer 7.

Several load balancing services are offered by Azure.

- **Azure Front Door.** Azure Front Door provides global load balancing for web applications. It provides Layer 7 functions including SSL offload, path-based routing, fast failover, and caching. It is a global load balancer used for HTTP(S) traffic.
- **Azure Traffic Manager.** Azure Traffic Manager is a DNS-based load balancer that distributes traffic across regions. It is a global load balancer used for non-HTTP(S) traffic.
- **Azure Application Gateway** Azure Applications Gateway provides various OSI Layer 7 load balancing capabilities including offloading SSL terminations to the gateway. It is a regional load balancer used for HTTP(S) traffic.
- **Azure Load Balancer.** Azure load balancer operates at OSI Layer 4 and is used for all TCP and UDP traffic. It is highly available and can scale to millions of requests per second. It is a global, non-HTTP(S) load balancer.

7.2.3.3   Autoscaling

When demand increases, virtual machines can be vertically or horizontally scaled. Vertical scaling increases the capabilities of an instance. For example, adding more CPU capacity to a single existing instance.

Horizontal scaling is growing by adding more instances. In other words, vertical scaling makes a virtual machine bigger, whereas horizontal scaling creates more virtual machines.

Horizontal autoscaling of virtual machines is accomplished in Azure Autoscale using VM Scale Sets. VM Scale Sets manage a group of virtual machines as a single group.

Azure Autoscale

7.2.4 Compute

As with other public cloud providers, the compute function is accomplished using virtual machines. Azure actually named its compute function the Azure "Virtual Machine."

7.2.4.1 Virtual Machine

Azure compute virtual machine capability is simply called "Virtual Machine." Azure Virtual Machine types include the operating system (Linux or Microsoft Windows), the number of virtual CPUs, the amount of memory and storage, and the processor architecture (Intel, AMD, or ARM). Azure can deploy virtual machines with up to 416 vCPU and 12 TB of memory.

Azure Virtual Machine

Azure provides a wide variety of Virtual Machines to meet just about any computing needs in the following categories:

- **General purpose.** Azure general-purpose Virtual Machines provide a balanced CPU-to-memory ratio. General purpose Virtual Machines are ideal for testing and development, small to medium databases, and low- to medium-traffic web servers.
- **Compute optimized.** Azure compute-optimized Virtual Machines are designed with a high CPU-to-memory ratio. Compute-optimized Virtual Machines can be used for medium-volume web traffic and application servers.
- **Memory optimized.** Azure memory-optimized Virtual Machines are designed with a high memory-to-CPU ratio. Memory-optimized instances are a good match for relational databases.
- **Storage optimized.** Azure storage-optimized Virtual Machines provide high disk throughput and are ideal for SQL and NoSQL databases and data warehouses.

- **High-performance compute.** Azure high-performance compute Virtual Machines have fast and powerful CPU virtual machines with optional high-throughput network interfaces (RDMA).

Reserved Virtual Machines

Azure Reservations are available for Azure Virtual Machines. For a one- or three-year commitment, the Reservations service can reduce the cost of Virtual Machines by 70% or more. If the usage of your virtual machines is steady and predictable, Reservations is a good choice over on-demand instance.

Spot Virtual Machines

Azure Spot Virtual Machines allow the customer to use excess Virtual Machine capacity in Azure when it is available for huge cost savings. However, spot capacity availability is unpredictable and fluctuating. If Azure needs the spot capacity, you are using they can reclaim it with only a 30-second notice.

Spot Virtual Machines are ideal for workloads that are lower priority or can be interrupted without detriment to the workload process. If so, Spot Virtual Machines can reduce virtual machine costs by 90% or more over on-demand instances (Microsoft, 2023a).

7.2.5   Storage

Azure storage services include ephemeral, block, file, and object storage options.

7.2.5.1   Azure Ephemeral Storage

Ephemeral, or short-lived, storage can be created on Azure Virtual Machines. Ephemeral storage is "non-persistent," which means if the Virtual Machine is stopped, the data in ephemeral storage is lost.

It may not sound like using ephemeral storage should ever be used. However, it is ideal for "stateless" operations where the data does not have to be saved after it is used (Microsoft, 2023b). An example is a website shopping cart. Once the order is paid for, the data is moved to a database and is no longer needed in the ephemeral storage.

7.2.5.2 Azure Files

Azure Files is a file storage service that allows for the creation of conventionally formatted file shares that are accessible using SMB and NFS protocols for Windows, Linux, and MacOS. Using Azure File Sync local and Azure Files can have the same files resident simultaneously.

Azure Files

Currently, Azure Files supports locally redundant storage (LRS), zone redundant storage (ZRS), geo-redundant storage (GRS), and geo-zone-redundant storage (GZRS).

- *Locally redundant storage (LRS)* copies data three times within a single physical location in the primary region.
- *Zone-redundant storage (ZRS)* copies data across three Azure availability zones in the primary region.
- *Geo-redundant storage (GRS)* copies data three times to a single physical location in the primary region using LRS, then copies that data to a single physical location in a secondary region that is hundreds of miles away from the primary region. GRS offers durability of at least 99.99999999999999% (sixteen 9's) over a given year.
- *Geo-zone-redundant storage (GZRS)* combines the high availability provided by redundancy across availability zones with protection from regional outages provided by geo-replication. Data is copied across three Azure availability zones in the primary region and also replicated to a secondary geographic region. GZRS also offers durability of at least 99.99999999999999% (sixteen 9's) over a given year.

7.2.5.3 Azure Disk Storage

Azure Disk Storage is a durable, low-latency block storage service for Azure Virtual Machines. It is implemented using older hard disk drives (HDD) and newer and faster solid-state drives (SSD).

Azure Disk Storage

Different levels of Azure Disk Storage performance are available:

- *Ultra Disk Storage* is used when sub-millisecond latency is required. Premium SSDs are suitable for mission-critical production applications.
- *Premium SSD* is available when production workloads require low latency and high-performance storage for mission-critical applications.
- *Standard SSD* provides cost-effectiveness, consistent low latency, and good performance for many production workloads. Standard SSDs are suitable for web servers and other basic applications.
- *Standard HDD* is appropriate for non-critical workloads, backups, and infrequently accessed data. Azure standard HDDs deliver reliable, low-cost disk support for VMs running latency-insensitive workloads.

Azure Storage Redundancy in Azure Disk Storage is accomplished in two ways:

- *Locally redundant storage (LRS)* copies data three times within a single physical location in the primary region.
- *Zone-redundant storage (ZRS)* copies data across three Azure availability zones in the primary region.

7.2.5.4   Azure BLOB Storage

Azure BLOB storage is Azure's massively scalable and secure object storage service. "BLOB" stands for "Binary Large Object." Azure BLOB storage can store virtually unlimited amounts of unstructured data.

Azure BLOB Storage

Azure BLOB Storage has a durability of 99.999999999%, or "eleven 9's." Durability refers to the likelihood, out of 100%, that your data will be available to you when you want to access it. This is the equivalent of the chances of losing only one of 10,000 files over a 10,000,000-year period!

High durability is accomplished by Azure BLOB Storage using various redundancy methods. All Azure files can use **locally redundant (LRS)** or **zone-redundant storage (ZRS)**. **Geo-redundant (GRS), geo-zone-redundant storage (GZRS)** are available for standard file shares under 5 TB.

Azure BLOBs have three tiers that are priced according to how the data is accessed. *Hot storage* is the most expensive, but you have immediate access to your data. Storing data here is the most expensive, but accessing it is the least costly. *Cool storage* is for data that is accessed less often but must be quickly available if need be. There is an additional cost for cool storage if data is not kept cool for thirty days.

Archive storage is for data that is to be stored but seldom, if ever, accessed. Storing data here is cheap, but accessing it is costly. Retrieving data from the archive can take up to 15 hours. Data must stay in the archive for 180 days, or it will incur more charges.

Azure administrators can set parameters for automatically moving objects from hot, cold, and archive storage such that storage costs can be minimized.

7.2.6   Databases

Microsoft Azure offers both relational and non-relational database services.

7.2.6.1   Relational Database

Azure SQL Server is Azure's relational database service.

Azure SQL Server

Microsoft's SQL Server relational database adapted to the cloud is called Azure SQL Server. It is a PaaS service.

Azure SQL is a good choice for a fully managed service solution, which means Azure takes care of all maintenance and upgrades, if desired. If high reliability is a required Azure SQL replicates data across availability zones. Azure SQL will also automatically scale to meet any demand.

Azure SQL Server

Azure SQL can be deployed either as a single database or in an elastic pool of single databases with shared resources. Elastic pools provide a cost-effective solution for managing the performance of multiple databases that have variable usage patterns.

Azure SQL Managed Instance

If complete access to the database engine is required, Azure allows deploying a virtual machine with an image of SQL Server built in to allow more user control and guarantees an always-up-to-date SQL instance in the cloud. However, if SQL Server on a VM is used, maintenance responsibility reverts back to the user.

Azure SQL Server on Azure Virtual Machines

SQL workloads can be migrated to Azure while maintaining system-level access.

Azure Database for PostgreSQL

Migration of PostgreSQL or Oracle workloads to Azure.

Azure Database for MySQL or MariaDB

Migration of emobile and web apps with MySQL
    Azure SQL offers three service tiers:

- *General purpose/standard service:* Based on separation of compute and storage. Highly available and reliable due to the replication of data.
- *Business-critical/premium service:* Based on database cluster processes.
- *Hyperscale service:* Highly scalable compute and storage performance.

7.2.6.2   Non-relational Database

CosmoDB is Azure's non-relational database service.

CosmoDB

CosmoDB is Microsoft's Azure's proprietary cloud-based platform that can host many non-relational databases (Cassandra, MongoDB, and other NoSQL workloads) with guaranteed low latency, highly available and scalable distributed database. Specifically, CosmoDB characteristics include:

- Global distribution – transparently distributed across all chosen regions.
- Regional presence – available in all Azure regions.
- Availability – 99.999% for reads and writes.
- Elasticity – up to millions of requests per second.
- Low latency – less than 10 millisecond reads and writes.
- No schema or index management – NoSQL database.

Azure Cosmo DB

7.2.7   Azure Lab

You are now ready to experience Azure using the Azure Free Trial. Azure allows twelve free months of all major services. This is more than enough time to learn Azure and decide if you want to continue using it on a pay-as-you-go basis.

After signing up for an Azure Free Trial account, you will then create a basic Azure infrastructure that applies what you have learned up to this point (Fig. 7.2). Similar to the AWS lab, your design is a basic fully operational three-tier web application running on the Azure platform using the following services:

- VNet.
- Azure VM Scale Sets.
- Cloud Load Balancer.
- NAT.
- Azure SQL.
- Storage Account.
- Resource Group.

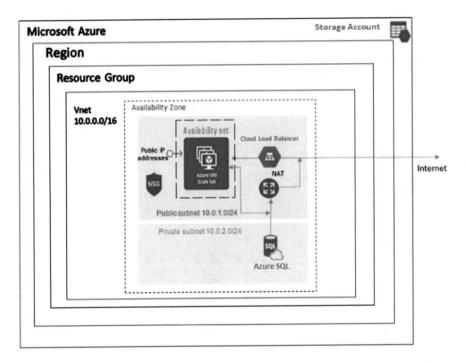

Fig. 7.2 Azure VNet

7.3   Summary

Microsoft Azure has a global infrastructure that spans geographies, regions, and availability zones and supports over 200 services with more being developed regularly.

Learning all the services offered by AWS would be very difficult, so this focus was on the most necessary and used services including:

- Monitor.
- Identity.
- Networking.
- Compute.
- Storage.
- Database.

Homework Problems and Questions

1.1 Describe the Microsoft Azure global infrastructure and its related components.

1.2 Describe the following Azure services:

- Azure monitor.
- Azure Active Directory.
- Azure VNet.
- Azure Load Balancing.
- Azure Autoscaling.
- Azure Files.
- Azure Disk Storage.
- Azure BLOB Storage.
- Azure SQL Server.
- Azure CosmoDB.

Bibliography

Amazon. (2023). *Azure monitor pricing*. [Online]. Available at: https://azure.microsoft.com/en-us/pricing/details/monitor/. Accessed 11 Jan 2023.

Azure. (2023). *Azure global network*. [Online]. Available at: https://azure.microsoft.com/en-us/explore/global-infrastructure/global-network/#overview

Microsft Azure. (2023). *Azure products*. [Online]. Available at: https://azure.microsoft.com/en-us/products/. Accessed Jan 2023.

Microsoft. (2023a). *Azure spot virtual machines*. [Online]. Available at: https://azure.microsoft.com/en-us/products/virtual-machines/spot/#overview. Accessed 16 Jan 2023.

Microsoft. (2023b). *Ephemeral OS disks for Azure VMs*. [Online]. Available at: https://learn.microsoft.com/en-us/azure/virtual-machines/ephemeral-os-disks. Accessed 15 Jan 2023.

Chapter 8
Cloud Platform

8.1 Google Cloud Platform (GCP) Global Infrastructure

The Google Cloud Platform services are available globally in over 200 countries. Components of the Google Cloud network infrastructure include (Google Cloud Infrastructure, 2023):

- 35 regions
- 106 zones
- 27 data centers

  - North America: 15
  - South America: 1
  - Europe: 6
  - Asia: 2

- 176 network edge locations

The GCP infrastructure is supported by regions, availability zones (AZs), edge locations, and local zones.

8.1.1 GCP Regions and Zones

Regions are usually paired with three availability zones (Table 8.1). An example of regions with associated zones is below. Zones provide redundancy capabilities within regions.

Not all services are available in all regions. Prices for the same services in different regions may be priced differently.

© The Author(s), under exclusive license to Springer Nature Switzerland AG 2024   143
M. S. Kingsley, *Cloud Technologies and Services*, Textbooks in
Telecommunication Engineering, https://doi.org/10.1007/978-3-031-33669-0_8

Table 8.1 GCP region examples (Economize, 2023)

Region	Location	Zones
Southamerica-east1	São Paulo, Brazil	Southamerica-east1-a
		Southamerica-east1-b
		Southamerica-east1-c
Europe-west2	London, U.K.	Europe-west2-a
		Europe-west2-b
		Europe-west2-c
Asia-south1	Mumbai, India	Asia-south1-a
		Asia-south1-b
		Asia-south1-c
Australia-southeast1	Sydney, Australia	Australia-southeast1-a
		Australia-southeast1-b
		Australia-southeast1-c

8.1.2   GCP Edge Nodes

Google Edge Nodes support the Google Global Cache system, which provides caching for static content close to customers.

8.1.3   Google Network

All Google services, including the Google Cloud Platform, are supported by a Google-owned global fiber optic network that is designed for resiliency and speed (Fig. 8.1). According to some estimates, the Google Fiber network carries 25% of the world's Internet traffic.

8.1.3.1   Google Network Service Tiers

Network services are available from Google in two tiers: standard and premium.

- *Standard tier network service* is used to optimize costs. In general, customer traffic is routed over the public Internet.
- *Premium tier network service* is a unique service to Google Cloud. It is used to optimize performance. Customer traffic is routed as much as possible over the Google private network. Recommended when service needs to be globally available. Premium tier is the default network service unless standard is manually selected.

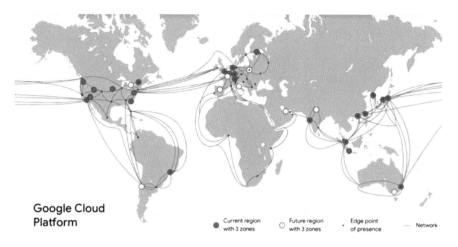

Fig. 8.1   Google network. (Google Cloud Network, 2023)

8.2   GCP Services

GCP offers more than 200 services in the following categories:

- IAM and Admin
- Compliance
- Security
- Anthos
- Compute
- Serverless
- Storage
- Databases
- Networking
- Application integration
- Databases
- Operations
- CI/CD
- Big data
- Artificial intelligence
- Partner solutions

It would be difficult and time-consuming to describe all of Google's services. We will look at a reduced number of the services necessary to create an operational three-tier web architecture: Selected services in the following categories will be covered:

- Management tools
- Identity and security
- Networking
- Compute
- Storage and database

8.2.1 Management Tools

The primary tools needed to configure and manage your Google infrastructure are the Google Cloud Console and Operations Suite.

8.2.1.1 Cloud Console

The Cloud Console is the web-based portal that interfaces the customer to the Google Cloud Platform.

**Google Cloud
Console**

8.2.1.2 Operations Suite (Formerly Stackdriver)

Google Cloud's operations suite (formerly Stackdriver) is software for monitoring and managing GCP networks and systems. It includes the following components:

- *Cloud Monitor:* Extensive monitoring for Google Cloud Platform.
- *Cloud Logging:* Records maintenance activity and performs various analyses.
- *Error Reporting:* Identifies errors in GCP services.
- *Cloud Debugger:* Identifies coding errors.
- *Cloud Trace:* Collects and displays application latency data.
- *Cloud Profiler:* Allows developers to analyze running applications.

**Google Operations
Suite**

8.2.2 Identity and Security

Basic Google Cloud Platform user identity, authorization, and authentication are provided by Cloud IAM.

8.2.2.1 Cloud IAM

Cloud IAM is GCP Identity and Access Management (IAM). It allows the assignment of user privileges by defining who has what access capabilities for which GCP resources.

Cloud IAM

8.2.3 Compute

The compute function is accomplished by the Google Cloud Platform's Compute Engine.

8.2.3.1 Compute Engine

Google Cloud Engine enables running applications on virtual machines, or instances, on physical hardware in Google's global data center. Cloud Engine runs on Linux or Windows operating systems and can be defined with varying CPU, memory, and storage capabilities by using predefined machine types.

**Google Compute
Engine**

Predefined Compute Engine machine types include the following:

- *General purpose:* Best balanced combination of compute and price performance. Good for smaller web servers and databases.

- *Compute-optimized:* Highest per-core performance for compute-intensive work-loads such as gaming.
- *Memory-optimized:* More memory per core, up to 12 TB. Applications include large databases and analytics.
- *Accelerator-optimized:* Used for workloads such as machine learning (ML) and high performance computing (HPC) or for applications that require GPUs.

Custom machine types can be created when exact compute and memory specifications are desired.

8.2.4   Networking

Networking components of the Google Cloud Platform include the virtual private cloud, load balancing, and autoscaling.

8.2.4.1   Virtual Private Cloud (VPC)

A VPC is a logically isolated and secure partition of the Google Private Cloud network that provides customer with access to GCP resources. A VPC is a virtual, or software-defined, representation of a physical network. A VPC contains both public and private subnets. Users provision compute, storage, databases, and other virtual machine resources inside the VPC.

**Google Virtual Private Cloud
(VPC)**

A GPC customer is not aware of the specific physical location of their virtual resources. They can be geographically separated and span multiple zones and regions.

8.2.4.2   Cloud Load Balancing

Load balancing enables an equal distribution of traffic to resources. GCP provides several load-balancing options.

The Google Cloud Platform has two basic options for load balancing. An ***external load balancer*** balances traffic coming from the Internet to publicly accessible Google Virtual Private Cloud (VPC) instances. ***Internal load balancers*** distribute traffic inside the Google Cloud Platform to GCP instances.

Global load balancers operate on customer resources that span different regions. Google's global load balancers operate differently than most DNS-based load balancers in that they are located at the edges of regions, resulting in more efficient load balancing.

Regional load balancing is used when customer resources are confined to one region.

Global load balancing requires premium tier network service.

Load balancer

8.2.4.3   Autoscaling

Autoscaling is accomplished on the Google Cloud Platform using Scale Sets. It increases or decreases the number of compute engines as demand changes. It is accomplished by creating a managed instance group of the desired number of compute engines included in the autoscaler. Autoscaling groups can be single- or multi-zoned.

Google Scale Sets

Autoscaling is determined by setting an autoscaling policy and setting utilization levels of instance parameters such as CPU usage and other monitored metrics. Once metrics are determined to be out-of-bounds, instances will be triggered to turn on or off to meet the increased or decreased demand. Charges are not applied when an instance is not turned on, allowing for granular adaption to demand rather than enabling instances for the maximum demand and resulting cost continuously.

8.2.5   Storage

GCP defines several types of storage: ephemeral, file, block, and object storage.

8.2.5.1   Local SSD

Local ("ephemeral") solid state drive (SSD) storage is physically attached to a physical server that hosts Compute Engine instances. If the instance is stopped or deleted, the stored data is lost. A physical server can have a total of 24 local SSD partitions of 375 GB, each for a total of 9 TB. Local SSD can operate using either SCSI or NVMe protocols.

8.2.5.2   Cloud Filestore

Google Cloud Filestore brings the simplicity of NFS-based Network Attached Storage (NAS) to the GCP. It is a high performance, fully managed shared file storage system for Google Compute Engine and Kubernetes Engine instances. It provides low latency, high IOPS, and predictable pricing. Performance is consistent and can be fine-tuned by adjusting IOPS. Filestore files can be accessed from either a Compute Engine instance within the same VPC or remote clients.

Google Cloud Filestore

Filestore High Scale enables the deployment of shared file systems that can scale out to high IOPS, throughput, and storage. Filestore is a good fit for applications such as data analytics, web content management, and media-intensive applications.

8.2.5.3   Persistent Disk

Persistent disk is block storage and can be attached to Compute Engine instances. "Persistence" refers to the capability of stored data to exist when separated from the Compute Engine instance.

The Google Cloud Platform has two persistent disk options. Zonal persistent disks reside in a single one zone; regional persistent disks are similar but provide data replication between two zones in the same region.

Google Persistent Disks

There are four persistent disk options available for zonal persistent disk implementations, depending on cost and performance objectives desired. Data redundancy is limited to a single zone.

- *Standard Persistent Disks (pd-standard)* are efficient, reliable, and economical block storage. Standard PDs are delivered on standard hard disk drives. They are appropriate for cost sensitive applications or cold storage.
- *Balanced Persistent Disks (pd-balanced)* are designed for a balanced cost-performance ratio. They are a good fit for most workloads including most enterprise applications and as boot drives. Balanced PDs are delivered on solid state drives (SSD).
- *SSD Persistent Disks (pd-ssd)* are fast and reliable block storage. SSD PDs are provided on solid state drives (SSD). They can be used for performance-sensitive workloads including most databases, persistent caches, and scale-out analytics. SSD persistent disks are delivered using solid state drives (SSD).
- *Extreme Persistent Disks (pd-extreme)* are chosen if extreme disk performance is required. Extreme PDs are intended primarily for use with high-end databases. Extreme persistent disks are delivered using solid state drives (SSD).

Regional persistent disk operation includes synchronous replication of data across two zones in a region of standard, balanced, SSD, and extreme persistent disks.

8.2.5.4 Cloud Storage

Cloud Storage is an object storing service in the Google Cloud Platform. Objects can be data of any format and are stored in "buckets." All buckets belong to a "project" and projects can be grouped in "organizations."

A Cloud Storage bucket can host a static website. Objects in a bucket can, if allowed, be accessed from anywhere on the Internet. Cloud Storage Engine scales to almost unlimited capacity.

Cloud Storage

Linux, Windows, as well as private custom images that you can create or import from your existing systems are supported.

When creating a Cloud Storage bucket, the location where the object will be stored has to be defined. Location choices are:

- A region.
- Dual region is a defined pair of regions.
- A multi-region is a large geographic area, such as the United States, that contains two or more geographic locations.
- Multi-region and multi-region objects are *geo-redundant*.

Cloud Storage supports four storage classes:

- *Standard storage:* Appropriate for "hot" data that is accessed often or for data that is not stored for very long. Website, streaming video, and mobile applications are a good fit for standard storage.
- *Nearline storage:* Highly durable storage for data that can be stored for at least 30 days. Retrieving data before 30 days will incur extra cost. Nearline is lower-cost storage, but retrieving data cost is higher. Ideal for data that is read or modified once per month or less. Appropriate for data backups and data archiving.
- *Coldline storage:* Very low-cost storage for data that is seldom accessed. However, data retrieval cost is expensive. Appropriate for data that can be stored for at least 90 days. Retrieving data before 90 days will incur extra cost.
- *Archive storage:* Lowest cost storage and highest retrieval cost. Data must be stored for at least 365 days. Removing data before 365 days incurs more cost.

Google Cloud storage allows cost reductions by moving objects to Nearline and Coldline and through scheduled deletions.

8.2.6   Databases

Google Cloud Platform offers two relational and two non-relational database services. There is also the option to run an Oracle database on a bare metal server.

8.2.6.1   Relational Databases

GCP relational databases are Cloud SQL and Cloud Spanner.

Cloud SQL

Cloud SQL is a globally accessible, fully managed relational database service for MySQL, PostgreSQL, and SQL Server. Backups and data replication are easily automated, and capacity increases are automatic. Cloud SQL can be configured for high availability.

Cloud SQL uses include ERP, CRM, E-commerce, web, and SaaS applications.

Cloud Spanner

Relational databases typically do not scale well horizontally. Cloud Spanner is a managed relational database service designed for unlimited scale, strong consistency, and high availability. It can service millions of requests per second and creates three read-write replicas, each in a different zone. Cloud Spanner still allows the use of SQL queries.

Google Cloud Spanner

8.2.6.2   Non-relational Databases

GCP non-relational databases are Cloud BigTable and Datastore.

Cloud Bigtable

Cloud BigTable is a fully managed, highly scalable, cloud-native NoSQL database service for large-scale applications with low latency requirements. It can handle millions of requests per second with less than ten milliseconds of latency. Clusters can scale to hundreds of nodes without downtime. Used by Google Search and Maps Cloud. BigTable integrates with big data tools like Hadoop and Dataflow and supports the open-source HBase API.

Cloud BigTable

Datastore

Datastore is a highly scalable NoSQL database for web and mobile applications.

Google Datastore

Many organizations currently use Oracle relational databases for legacy applications. This can be a hindrance when desiring to move on-premises data center operations to a public cloud provider. Bare Metal Solutions allows moving existing Oracle licenses to the GCP, which also allows legacy applications to run in the cloud.

8.3 GCP Lab

You are now ready to experience GCP using the GCP Free Trial. GCP provides 3 months or $300 credit account access to new users. This should be enough time to try GCP and decide if you want to continue using it on a pay-as-you-go basis.

After signing up for a GCP Free Trial account, you will then create a basic GCP infrastructure that applies what you have learned up to this point (Fig. 8.2). Your design is a basic fully operational three-tier web application running on the GCP platform using the following services:

- **Virtual Private Cloud (VPC) network**
- **VPC firewall rules**
- **Cloud NAT**
- **Compute instance group**
- **Web servers**
- **Autoscaling**
- **Load balancer**
- **Cloud SQL database**
- **Cloud Storage bucket**

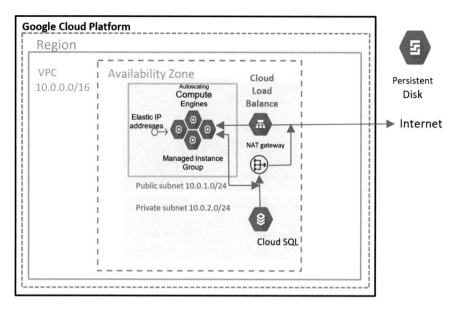

Fig. 8.2 GCP Lab

8.4 Summary

GCP has a global infrastructure that spans regions, zones, and edge locations and supports over 200 services with more being added regularly.

Learning all the services offered by AWS would be very difficult, so this focus was on the most necessary and used services including:

- Management tools
- Identity and security
- Networking
- Compute
- Storage and database

Homework Problems and Questions

1.1 Describe the GCP global infrastructure and its related components.
1.2 Describe the following service categories and the individual services they contain:

- Operations Suite
- Cloud IAM
- Virtual Private Cloud (VPC)
- Cloud Load Balancing
- Autoscaling
- Filestore

- Cloud Storage
- Persistent disk
- Cloud Storage
- Cloud SQL
- Cloud Spanner
- BigTable
- Datastore

Bibliography

Economize. (2023). *Regions and Zones.* [Online] Available at: https://www.economize.cloud/resources/regions-zones-map.
Google Cloud Infrastructure. (2023). *Google Cloud Infrastructure.* [Online] Available at: https://cloud.google.com/infrastructure.
Google Cloud Network. (2023). *Google improves connectivity with upgrades to its cloud infrastructure.* [Online] Available at: https://the-report.cloud/google-improves-connectivity-with-upgrades-to-its-cloud-infrastructure.

Part IV
AWS, AZURE, and GCP Labs

You have just finished an overview of the three largest public cloud providers:

- Amazon Web Services (AWS)
- Microsoft Azure
- Google Cloud Platform (GCP)

Now you will get the chance to use these platforms by completing a hands-on lab for each. You will create the same three-tier website architecture using the free tier services of each provider. The virtual private networks that you create for each provider are live and operational on each platform.

The following labs are waiting for you in this section:

- *Chapter 9: Amazon Web Services (AWS) Lab*
- *Chapter 10: Microsoft Azure Lab*
- *Chapter 11: Google Cloud Platform (GCP) Lab*

Here we go!

Chapter 9
Amazon Web Services (AWS) Lab

9.1 AWS Free Tier

You are now ready to experience AWS hands-on using the AWS Free Tier. The free tier offers 12 months of free service for dozens of AWS services. Some services have limits on free use. For example, AWS EC2 can be used for free for 750 h, which is more than enough time to practice using EC2. Other services have similar usage limits, while some services are free forever.

When you have an AWS account, you will then create a basic AWS infrastructure that applies what you have learned previously. Your design is a basic fully operational three-tier website configuration running on the AWS platform.

The first step is to create your AWS Free Tier account.

Setting up your AWS Free Tier Account
Begin by going to AWS Free Tier and select "Create a Free Account."

© The Author(s), under exclusive license to Springer Nature Switzerland AG 2024
M. S. Kingsley, *Cloud Technologies and Services*, Textbooks in
Telecommunication Engineering, https://doi.org/10.1007/978-3-031-33669-0_9

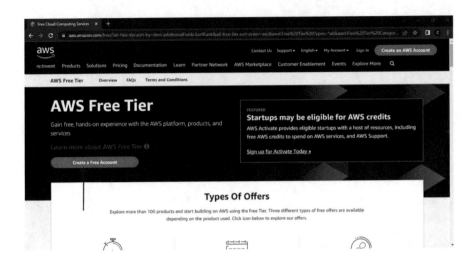

The signup process is straightforward. You will be asked to enter some basic information and provide credit card information to create an AWS username and password. There is no need to worry; AWS is very secure, and the credit card will only be used if you exceed your allotted free tier usage. How to avoid unintentional AWS charges will be described in later labs.

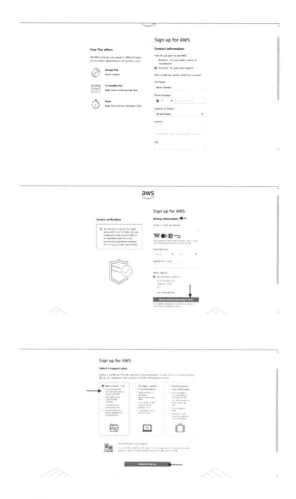

When completed you will receive a welcome email message from AWS confirming your access to the Free Tier.

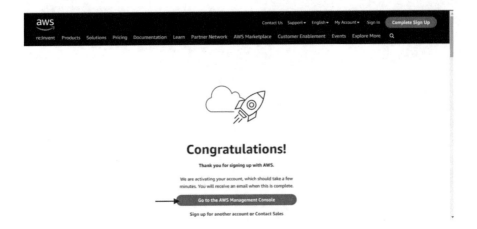

You are now ready to use AWS!

9.2 AWS Management Console

The AWS Management Console is the main dashboard that you will use to access and use different AWS services. AWS recently redesigned their user interface. We will use the new interface in this lab.

Use your AWS account email address and password to sign into the AWS Management Console as the AWS account root user.

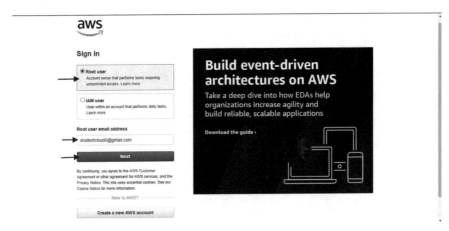

You are now ready to start building your AWS configuration! The first step is to use AWS Identity and Access Management (IAM) to create your user account security if desired.

9.2.1 AWS IAM

When you sign up for AWS, you are given "root" account access. Root access is very powerful and should never be used. Instead, you should create an administrator account for yourself using Identity and Access Management (IAM). Later you can assign access privileges to other users.

First, click on the **Services** icon in the upper left corner. Scroll down and select the **Security, Identity, & Compliance** link.

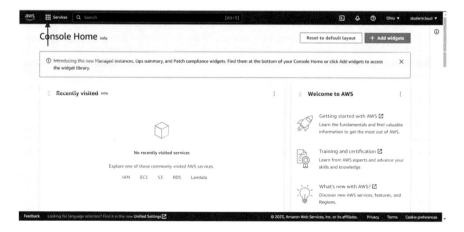

On the second screen to the right is the **Security, Identity, & Compliance**. Scroll down and click on **Identity and Access Management (IAM)** and follow these steps:

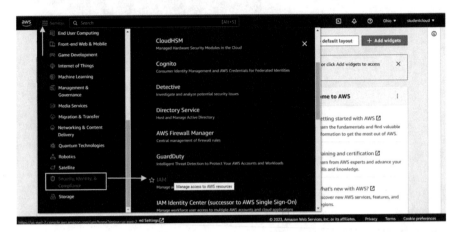

In the navigation pane of the console, choose **Users**, and then choose **Add user**s.

The **Add user** page will appear. Enter the following information:

Username: Administrator
Enable Console Access: enable
Console password: Custom password and then type the new user's password in the text box.
User must create a new password at next sign-in: Enable
Click on **Next**

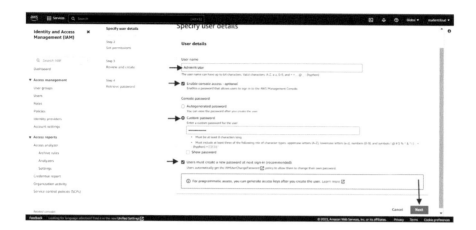

"Set Permissions" A new page will appear.
Under **permissions option**, choose to **Add user to group**.
Click on **Create group**.

The **"Create user group"** new page will appear.
User group name: Administrator
From the **Filter policies** drop-down menu, choose **AWS managed-job function**
and also search for "**IAMFullAccess**" and select it.

You will return to the "**Set Permission**" page. Select **Administrator** and click on **Next**. "**Review and Create**" page appears, click on **Create user.**

You will see that an **Administrator** has been added to the group. Now click on **Users(left pane)** and then **Administrator.** The **Summary** page will appear. Click on the **Permissions tab** and verify that **IAMFullAccess** is present.

Choose **Groups**: You will see the administrator group that we created.

Choose **Tags – Add tags (optional).** IAM tags are key-value pairs you can add to your user. Tags can include user information, such as an email address, or can be descriptive, such as a job title. You can use the tags to organize, track, or control access for this user.

To sign in as this new IAM user, sign out of the AWS console and make sure to copy the AWS account ID from the URL highlighted in the above SS, where your_AWS_account_id is your AWS account number without the hyphens (for example, if your AWS account number is 1234-5678-9012, your AWS account ID is 123456789012):

You will be asked to change your password.

You should now be logged into the AWS Management Console. Check your region for "**Oregon.**"

You have now created your administrator access account and can start building your AWS configuration.

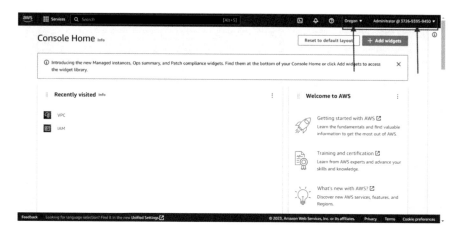

9.2.2   AWS Lab

In the AWS lab, a typical three-tier architecture will be produced by accomplishing the following tasks:

Task 1: Create and configure an AWS VPC, subnets, and an Internet gateway
Task 2: Create and configure VPC firewall rules
Task 3: Create and configure a NAT gateway and VPC route tables
Task 4: Create and configure AWS EC2 compute service
Task 5: Connect an elastic IP address
Task 6: Create and configure an autoscaling group
Task 7: Create and configure an Internet-facing Load Balancer
Task 8: Create and configure an AWS Aurora Database
Task 9: Create and configure an S3 storage bucket and connect it to EC2

Task 1: Create and Configure an AWS VPC, Subnets, and an Internet Gateway
Begin by creating your AWS infrastructure with a virtual private cloud (VPC). Public access, such as for web server requests, is accomplished using a public subnet and an Internet gateway; services that must be protected, such as databases, are deployed in private subnets. Separate services, in this case a database, from outside access with a private subnet. Allow web server access from a public subnet connected to an Internet gateway.

Task 1a: Create and Configure an AWS VPC
A virtual private cloud (VPC) is your logical slice of the AWS infrastructure. In other words, it is the box where most of your AWS services will be placed. It is

logical in that it may span many servers, even regions, of AWS. However, to you, it will seem as though everything is closely connected via a typical TCP/IP network. The VPC in the diagram below is highlighted in red.

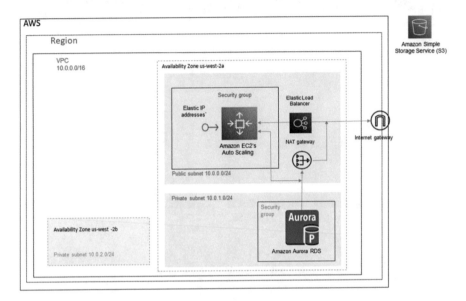

Click on the **Services icon** in the upper left-hand corner of the Management Console, then on **Network and Content Delivery**, and finally on **VPC** in the second menu that opens to the right.

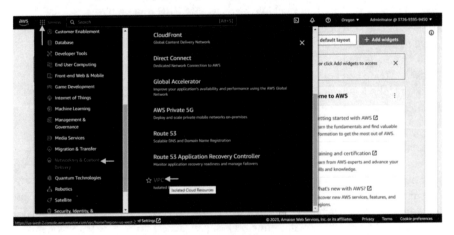

The **VPC Dashboard** page will appear. Click on **VPCs.**

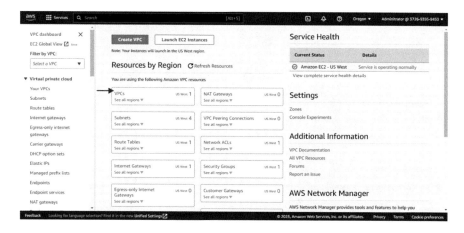

Notice that there is already a VPC there. This is the *default* VPC, and it cannot be deleted. You will create your own VPC. Click on **Create VPC.**

When the **Create VPC** page appears, enter the following information:

Resources to create: VPV only
Name tag: web-server-vpc
IPv4 CIDR block: 10.0.0.0/16
IPv6 CIDR block: no IPv6 CIDR block
Tenancy: Default

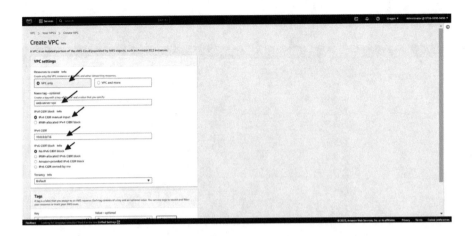

Scroll further down the page. The name tag should autofill the **Tags** field. If it does not, click **Add new tag.**

Key: Name
Value: web-server-vpc
Click **Create VPC**

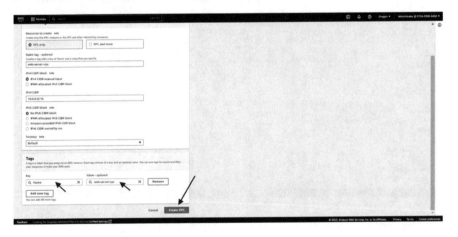

The next page will inform you that your VPC was created successfully. Click on **Your VPCs** (in the left-hand menu).

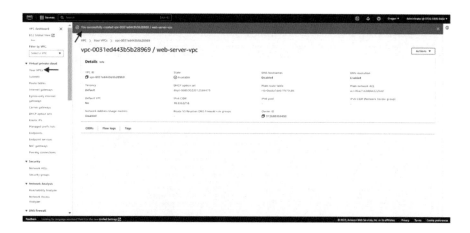

You should see that your web server VPC has been added along with the default VPC.

Task 1b: Creating and Configuring VPC Subnets

Next, you will add public and private subnets to your VPC. As their name implies, public subnets are accessible over the Internet, while private subnets are not. Services such as web servers are deployed in public subnets, while others such as databases that must be secured are deployed in private subnets.

As the diagram below illustrates, the public subnet is assigned an address space of 10.0.0.0/24, while the private subnet is assigned 10.0.1.0/24. Note that both subnets are part of the 10.0.0.0/16 space.

There is a second private subnet with an address space of 10.0.2.0/24 that will be added later in the lab.

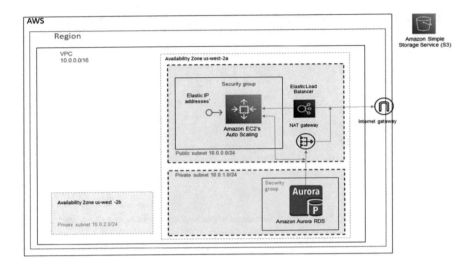

In the left menu, click on **Subnets**, and the Subnets page will appear. Click on **Create subnet** in the upper-right corner.

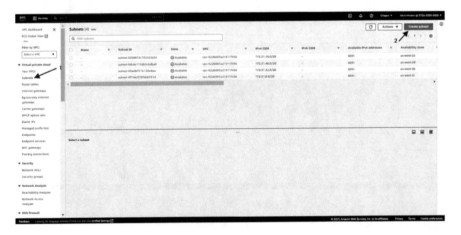

A public subnet will be created first.

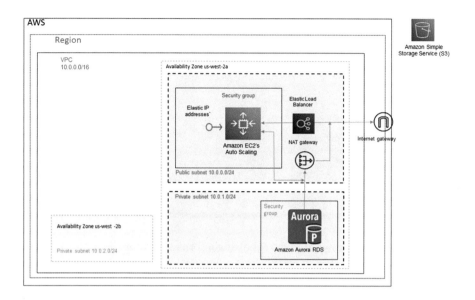

When the **Create subnet** page displays, add the following information from the drop-down menu:

VPC ID: vpc ######### (web-server-vpc)
Subnet name: PublicSubnet_1
Availability zone: No preference
IPv4 CIDR block: 10.0.0.0/24
Key: Name
Value: PublicSubnet_1 will autofill
Click Add New Subnet

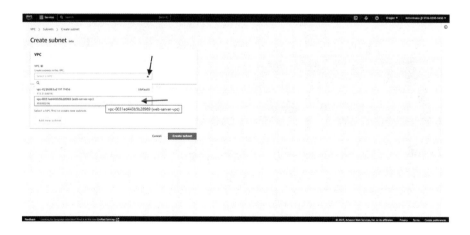

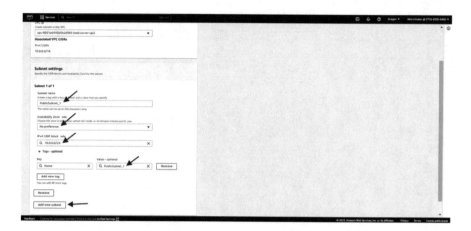

Scroll down on this page. Add the following information to the fields:

Subnet name: PrivateSubnet_1
Availability zone: No preference
IPv4 CIDR block: 10.0.1.0/24
Key: Name
Value: PrivateSubnet_1 will autofill.
Click Create subnet

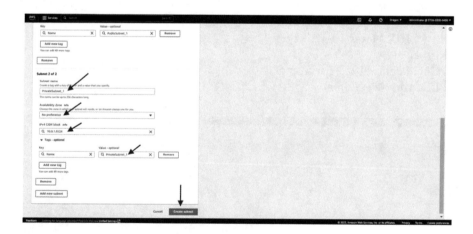

The **Subnet** page will appear. Both subnets should show they have been created and are available.

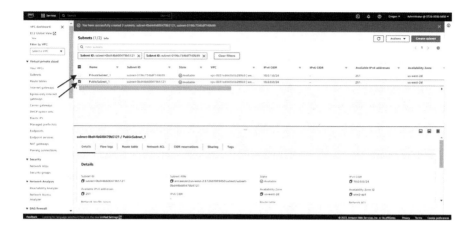

Task 1c: Create an Internet Gateway

An AWS Internet Gateway (IGW) provides access to the Internet.

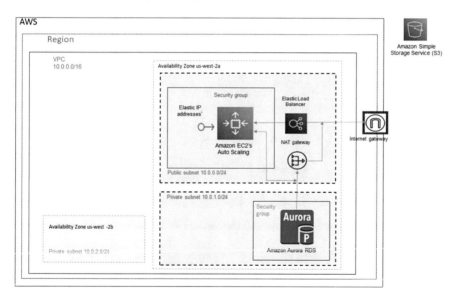

Click on **Internet Gateways** in the menu on the left under **Virtual private cloud** or enter "Internet gateway", in the search bar at the top of the page and click **Internet Gateways** in the drop-down menu.

When the **Internet gateways** pages appear, click on **Create internet gateway.**

The **Create internet gateway** page will appear.
Name tag: IGW_1
It will autofill in the **Tag** area. Then click on **Create internet gateway.**

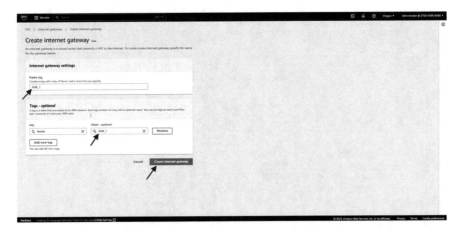

The page will refresh, indicating that the IGW has been successfully created but is "detached." Therefore, the next step is to attach it to the VPC you created. Click on **Attach to a VPC** at the top of the screen or from the drop-down **Actions** menu.

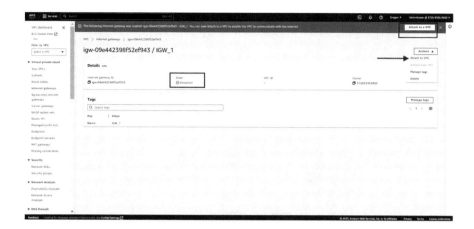

The **Attach to VPC** screen will appear. Click in the **Available VPC** field and select **web-server-vpc** previously created. Note: When the VPC is selected, only the VPC ID will fill in; the name information will not carry over. Click **Attach internet gateway.**

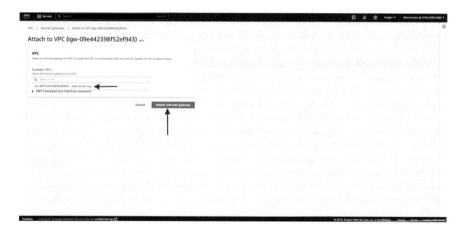

The screen will refresh, and the Internet Gateway will show "attached" to the VPC.

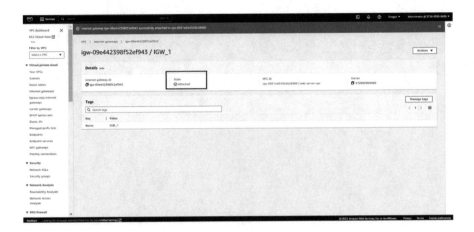

Task 2: Create and Configure VPC Firewall Rules

Under **Security** on the navigation menu, click on **Security Groups.** There are two *default* security groups: one for the default VPC and the other for the web-server-vpc you have created. (Note: **Name** field will be blank. Identify your VPC ID in "Your VPCs" and match it with below two VPC IDs on security group). Default security groups allow all incoming and outgoing VPC web traffic.

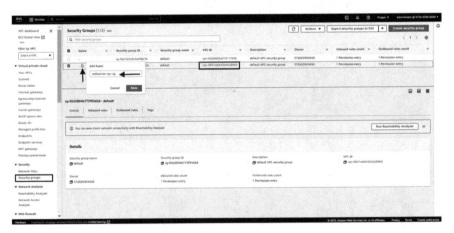

Traffic allowed into the web-server-vpc should be restricted. This is accomplished by creating a *custom* security group. Only HTTP and HTTPS **inbound** traffic will be allowed:

Protocol type	Protocol number	Port	Source IP	Notes
TCP	6	80 (HTTP)	0.0.0.0/0	Allows inbound HTTP access from any IPv4 address
TCP	6	443 (HTTPS)	0.0.0.0/0	Allows inbound HTTPS access from any IPv4 address

Select the web-server-vpc-sg by clicking in the check box, then on the **Security group ID** link.

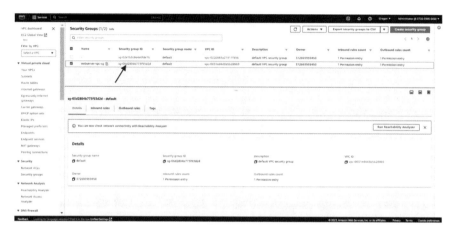

On the next screen, click on **Inbound rules** and then on **Edit inbound rules**.

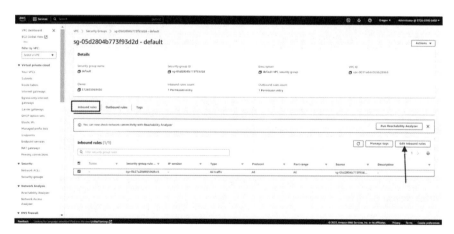

The **Edit inbound rules** page appears. Delete the "Allow all" rule.
Type: HTTP
Source: Custom

In the search field, select **web-server-vpc-sg**. In the **Description** field, enter "Allow inbound web traffic." Click on **Add Rule.**

Similarly, in the **Type** field, select "HTTPS" from the drop-down menu and "Custom" in the **Source** field. In the search field, select **web-server-vpc-sg**. In the **Description** field, enter "Allow inbound secure web traffic." Click on **Save rules.**

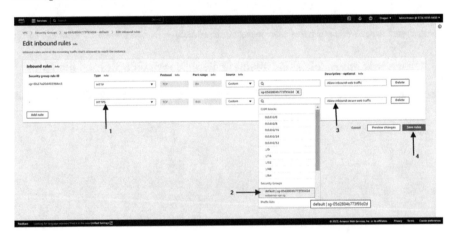

The page will refresh. Verify the **Inbound rules** now have HTTP and HTTPS rules applied.

Outbound traffic has no restrictions, so no outbound rules need to be created. Note: Any outbound traffic leaving the VPC will automatically be allowed to receive any replies.

Task 3: Create and Configure a NAT Gateway and VPC Route Tables

Users or services originating from the private subnet cannot access the Internet. However, there are cases where Internet access is needed. An example is when the database deployed in the private subnet needs to access the Internet to get database updates.

To accomplish, this requires the use of a *NAT gateway*. NAT connects a public IPv4 address to the private IP address of the database instance, which allows the database to access the Internet.

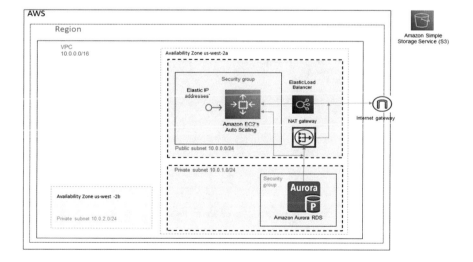

Task 3a: Create and Configure a NAT Gateway
Under **Virtual Private Cloud** in the navigation menu on the left side of the page, click on **NAT Gateways**. Click on **Create NAT gateway.**

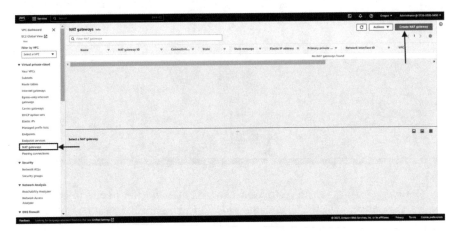

The **Create NAT gateway** page will appear. For **Name**, enter "NAT_1." NAT is deployed only in public subnets, so for **Subnet**, select PublicSubnet-1 of the web-server-VPC from the drop-down menu. Select **Connectivity type** as "Public."

A public IP address will be removed if an instance is stopped. An **Elastic IP address** is permanent. For **Elastic IP allocation ID**, select an elastic IP from the drop-down menu. If no elastic IPs are available, click on **Allocate Elastic IP** and then select that elastic IP from the drop-down menu.

Under **Tags**, "NAT_1" will autofill for the name. Click **Create NAT gateway.**

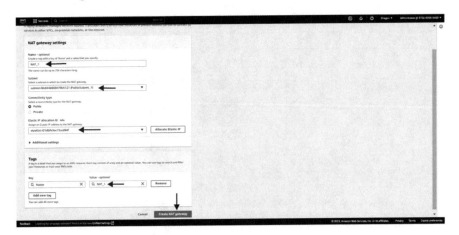

The NAT gateway details page will appear, verifying the NAT gateway with details. Select **NAT Gateways** from the navigation menu on the left of the page to monitor the progress of the NAT gateway creation and then the successful deployment.

Task 3b: Create and Configure VPC Route Tables

The web server VPC has both public and private subnets, but presently they cannot access the Internet. Each subnet has a default route table, which allows the subnets to access anything in the 10.0.0.0/16 address space. In other words, the public and private subnets can talk to each other. This is referred to as the "local" route. Any further routes require creating a custom route table.

Two custom route tables are needed. The first is for the public subnet to access the Internet via the Internet gateway. The second is for the private subnet to access the NAT gateway so the database in the private subnet can get updates. Route table fields include a **Destination** and a **Target.**

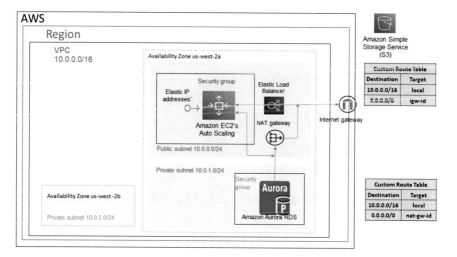

Both custom route tables have a "local" route, allowing each subnet to talk to the other. The custom route table for the private subnet now has a destination route for any IP address to the NAT gateway (nat-gw-id). The public subnet has a local route as well as a route to the Internet via the Internet gateway (igw-id) to any external IP address. The "id" number is a number assigned by AWS to uniquely identify the NAT or Internet gateway and is usually quite long.

Routes are defined by the route tables but are subject to the constraints imposed by security gateways, such as the VPC security gateway previously configured.

From the navigation pane on the left side of the page, under **Virtual Private Cloud,** click on **Route tables.** There are two default route tables, one for the web-server VPC you created and the other for the default VPC. Both contain local routes. We need to create two custom route tables, one for each subnet. Click **Create route table.**

Under **Name,** enter "PublicRouteTable_1." For VPC, select the web-server-vpc that we created. Name will automatically fill the **Tag** field. Click **Create route table.**

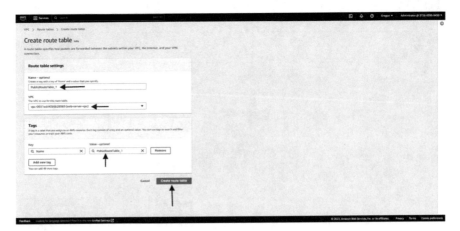

When the page refreshes, you will see the details of **PublicRouteTable_1**. Scroll down and click on **Subnet associations.** In the **Explicit subnet associations area**, click **Edit subnet associations**.

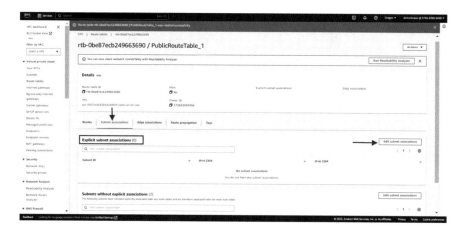

From the new page, select **PublicSubnet_1** and click **Save association.**

Go back to the **Route tables** page via the navigation pane. Select **PublicRouteTable_1** by clicking the check box. Click on **Routes** and then on **Edit routes.**

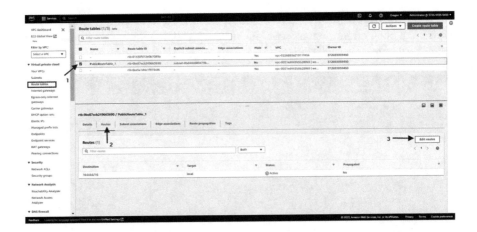

The **Edit routes** pages will appear. Select **Add route**, and for **Destination**, choose "0.0.0.0/0" from the drop-down menu. For **Target**, select the VPC Internet gateway "IGW_1." Click **Save changes.**

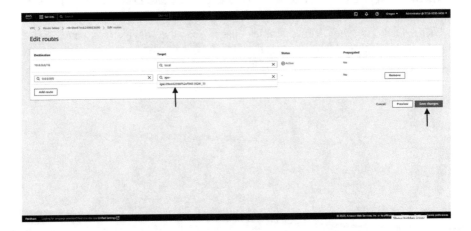

You have now created a route from the VPC public subnet to the Internet via the Internet gateway.

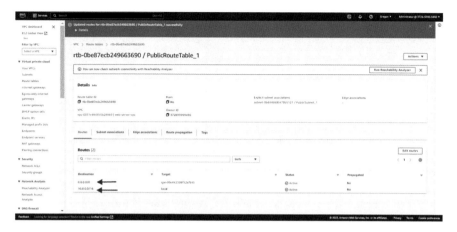

Following the same procedure, you now need to create a route from the private subnet to the NAT gateway.

From the navigation pane click **Route tables.** Click on **Create route table.**

The **Create route table** page appears. For **Name** enter "PrivateRouteTable_1". This will be copied into the **Tags value** field. For **VPC**, select the **web-server-vpc** from the drop-down menu.

Click on **Create route table.**

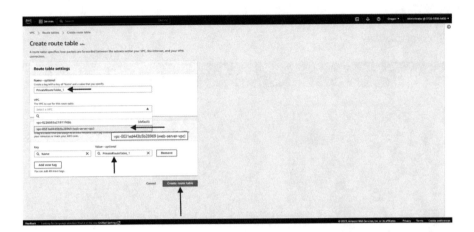

When the next page appears, click on **Subnet associations**, and then under **Explicit subnet associations**, click **Edit subnet associations**.

When the **Edit subnet associations** page appears, select the **PrivateSubnet_1** check box and then click on **Save associations**.

Return to **Route tables** from the navigation bar. Verify that the private route table has been created. Select **PrivateRouteTable_1** by selecting the check box. Click on **Routes** and then on **Edit routes.**

When the edit routes page appears, select **Add route.** For **Destination,** select "0.0.0.0/0" for **Target** and "NAT_1" from the drop-down menu. Select **Save changes.**

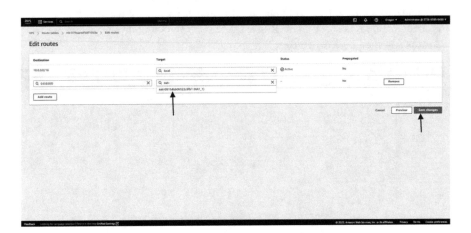

When the screen appears, it shows that the NAT route has been created. Return to the **Route tables** page from the navigation menu to see that **PrivateRouteTable_1** has been created in addition to **PublicRouteTable_1**.

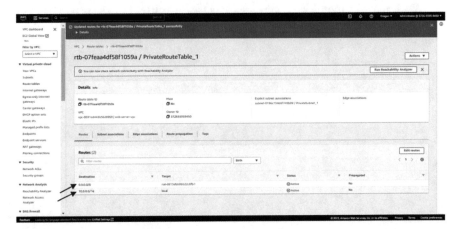

You have now created a route for the database in the private subnet to access the Internet via NAT to retrieve database updates when required.

Task 4: Create and Configure AWS EC2 Compute Service
Begin by creating and configuring a single EC2 with a web server application. Later, you will add more EC2s to enable autoscaling when traffic demand changes and load balancing to evenly distribute traffic between available EC2s.

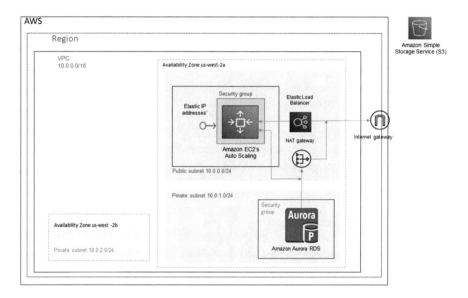

In the search bar at the top of the page, begin typing "EC2." Click on **EC2.**

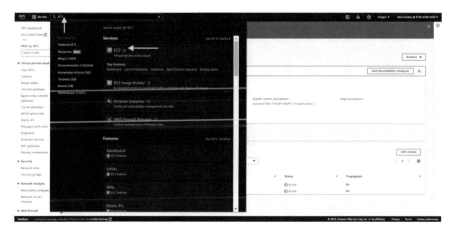

The EC2 dashboard appears (click on **EC2 Dashboard** if another screen appears). Notice that there are no instances running. Click on **Launch instance.**

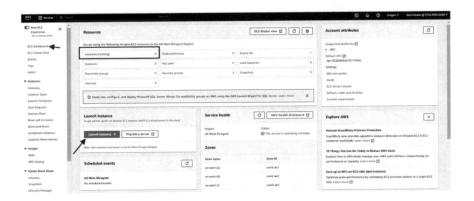

The **Launch an instance** page appears. Select **Name** as **WebServerEC2_1**

The **QuickStart** menu contains the most commonly used AMIs. You can also create your own AMI or select an AMI from the AWS Marketplace, an online store where you can sell or buy software that runs on AWS.

An AMI is a preconfigured arrangement of CPU, operating system, memory, and storage options available for EC2 instances. It also provides the information required to launch an instance including:

- A template for the root volume for the instance (for example, an operating system or an application server with applications)
- Launch permissions that control which AWS accounts can use the AMI to launch instances
- A block device mapping that specifies the volumes to attach to the instance when it is launched

Click **Select** next to **Amazon Linux 2 AMI**

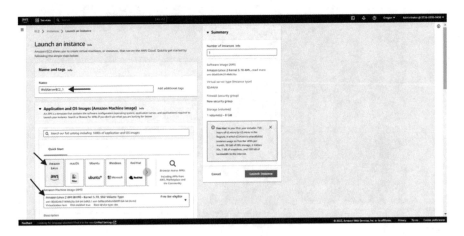

"Instance Type":

Amazon EC2 provides a wide selection of *instance types* optimized for different use cases. Instance types comprise varying combinations of CPU, memory, storage, and networking capacity and give you the flexibility to choose the appropriate mix of resources for your applications. Each instance type includes one or more *instance sizes,* allowing you to scale your resources to the requirements of your target workload.

Select **t3.micro** by searching under the drop-down menu. **t2.micro** provides a virtual CPU and 1 GiB of memory.

Choose an Key pain (login)

Select **Key pair name** as **Default.**

"Network Setting":

Enter the following information in the fields specified:

VPC: web-server-vpc

Subnet: PublicSubnet_1

Auto-assign IP addresses: Enable

Firewall(security groups): "Select an existing security group" and choose the "webserver-vpc-sg" that we created.

Inbound security group: "Remove"

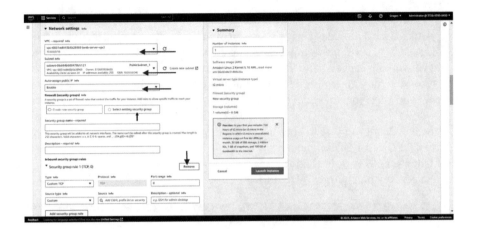

"Configure storage":

Amazon EC2 stores data on a network-attached virtual disk called *Elastic Block Storage*.

You will launch the Amazon EC2 instance using a default 8 GiB disk volume. This will be your root volume (also known as a 'boot' volume).

"Advanced details":

Hostname type: Resource name
DNS Hostname: Enable resource-based IPv4 (A record) DNS requests

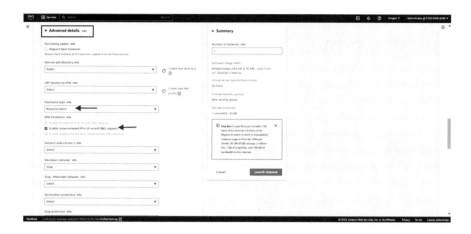

For **User data** paste in the following script:

```
#!/bin/bash
yum -y install httpd
systemctl enable httpd
systemctl start httpd
echo '<html><h1 > Hello From Your Web Server! </h1></html>'> /var./www/html/
   index.html
```

This script will create an Apache web server when this EC2 instance is created, as well as any created through autoscaling.

Select **number of instances as 1** on the right pane and click on **Launch instance**. Leave all other options as they are.

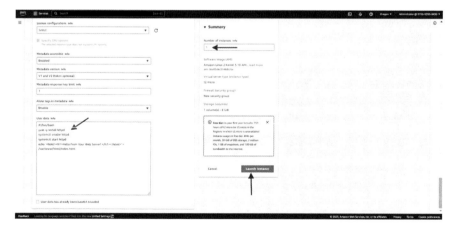

After this, the below screen will appear:

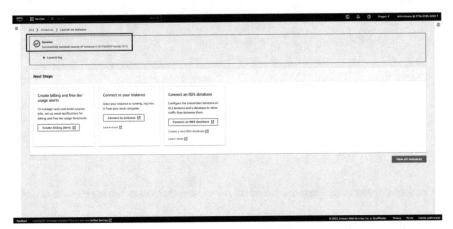

The next page shows that your EC2 launching is in progress. When finished, the status will show "Running," and the **Status check(s)** have passed.

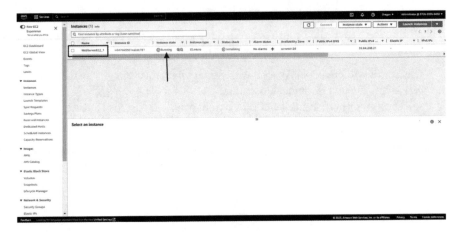

Below are images showing the details of EC2 that we created:

Task 5: Connect an Elastic IP Address

For the EC2 to connect to the Internet via the Internet Gateway. However, if the EC2 is stopped, a standard IP address will be released. The web server needs a permanent IP address called an elastic IP address.

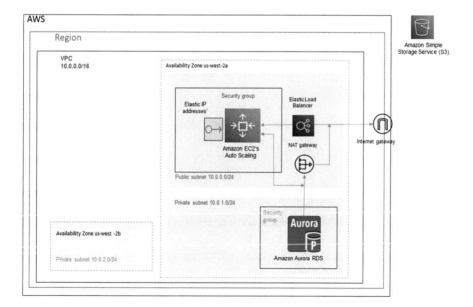

Under **Network and security** in the navigation menu, select **Elastic IPs**, and the related page appears. Select an available elastic IP by checking the related box. Under the **Actions** drop-down menu, select **Allocate Elastic IP address.**

From the **Allocate Elastic IP Address** page select:

Network Border Group: us-west-2
Public IPv4 address pool: Amazon's pool of IPv4 address
Click on **Allocate.**

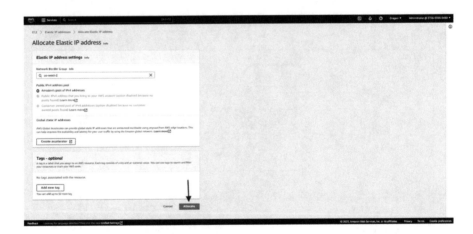

When the page appears, select the **Elastic IP address**, and under **Actions**, click on
Associate Elastic IP address.

Associate Elastic IP address page appears. Choose:

Resource type: Instance
Other details will be default. Click on **Associate**.

Go to the **Instance Summary for Web server EC2** and observe that the elastic IP address has been connected and is the private IP address used to access the web server.

Task 6: Create and Configure an Autoscaling Group

Autoscaling is used to increase or decrease the number of EC2 instances depending on fluctuations in web traffic. An autoscaling group can be created from scratch or from an existing EC2. Since you have already created the **Web server EC2**, autoscaling will be created from there.

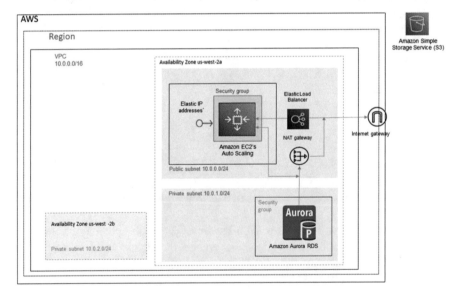

From the navigation pane, click on **Instances** and then click on **Web server EC2.** Under the **Actions** drop-down menu, select **Instance settings** and then **Attach to autoscaling group.**

The **Attach to Auto Scaling group** page appears. Under **Auto Scaling Group**, you have the choice to "Select an existing Auto Scaling group or enter a name to create a new Auto Scaling group." You will create a new Auto Scaling group by entering the name "ASG_1." Click on **Attach**, and a new Auto Scaling group launch configuration will be created.

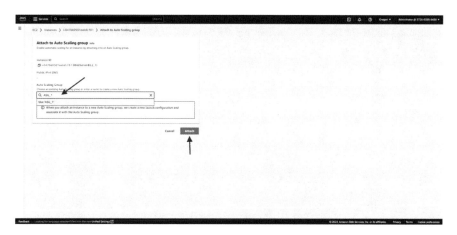

The **Instances** page will reappear. From the navigation, click on "Auto Scaling Groups" under **Auto Scaling.** The **Auto Scaling Groups** page will appear. Select ASG_1 and click on **Edit.**

The **Edit ASG_1** page appears. Enter the information below:

Desired capacity: Desired capacity is the number of EC2s that will initially be deployed: Set to "3"

Minimum capacity: Set to "1"

Maximum capacity: Set to "5"

Scroll down the page. **Launch configuration:** ASG_1

Under Network, **Available Zones and subnets:** PublicSubnet_1

Application, Network, and Gateway Load Balancer target groups: EC2 | HTTP

Note: "One of your groups is not yet associated with any load balancer. In order for routing and scaling to occur, you will need to attach the target group to a load balancer." You will do this later.

Health checks: ELB

Health check grace period: 10

Termination policies: Newest instance

Suspended processes: Terminate

Leave all other parameters as is.

Click **Update**

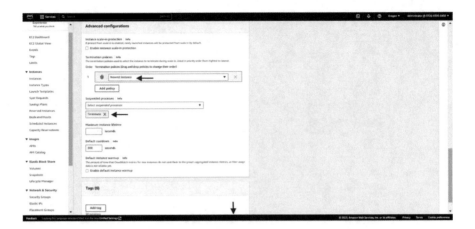

The **Auto Scaling groups** page appears. Notice there are three instances that have been created (the desired instance level).

Click on the **Instance management** tab to view the instance details. The **Auto Scaling groups** page appears. Verify that ASG_1 is being created. Click on the **Instance Management** tab. You should see three EC2s now running instead of just the single EC2 that you created earlier.

Scroll back up to view and click on the **Automatic Scaling** tab. **Dynamic scaling** reacts to changes in web traffic and scales out to add EC2s when needed and scales in to remove EC2s when web traffic is low. Click on **Create dynamic scaling policy.**

When the **Create dynamic scaling policy** page appears, provide the following information:

Policy type: Target tracking policy
Scaling policy name: Target tracking policy
Metric type: Average CPU utilization
Target value: 50

Others leave as it is. Click on **Create.**

The Auto Scaling Group will now attempt to keep the average EC2 instance CPU utilization at 50%. It should add instances if the average CPU utilization exceeds 50% and delete instances if it decreases to below 50%.

Task 7: Create and Configure an Internet-Facing Load Balancer
A load balancer will balance the traffic from the Internet among the instances provided by the Auto Scaling Group.

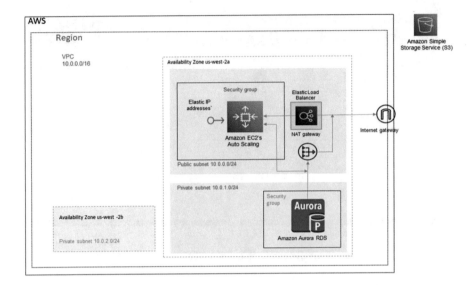

From the navigation menu, select **Load balancers** and then **Create load balancer.**

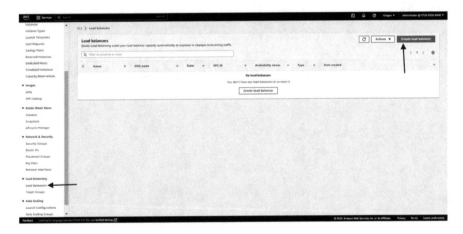

The **Select load balancer type** page will appear. Under **Application Load Balance** click on **Create.**

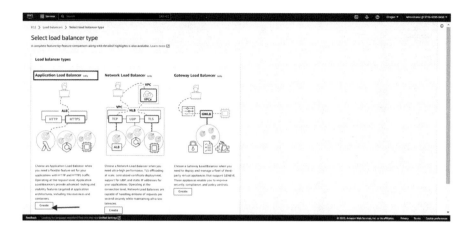

Create Application Load Balancer page will appear:

Under Basic configuration, choose:
Load balancer name: ALB_1
Scheme: enable Internet-facing
IP address type: enable IPv4

Scroll down on the **Network mapping** page and enter the following information:

VPC: web-server-vpc
Mappings: You need to choose two availability zones and one subnet per zone.
us-west-2a and select the available subnet
us-west-2b and select the available subnet

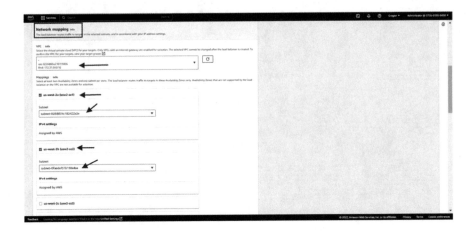

Security groups: Default
Listeners and routing leave "Listener HTTP:80" and **Create Target Group**

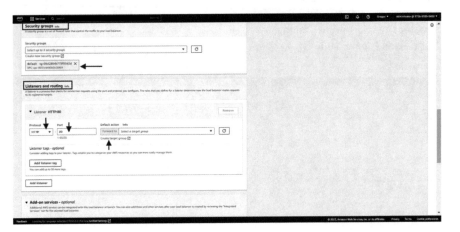

The **Basic configuration** page appears. Choose **Instances.** For **Target group name**, choose "EC2-TargetGroup-1" from the drop-down menu.

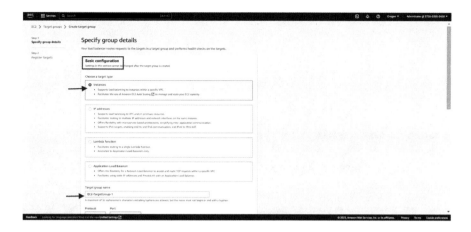

Scroll down to this page.
Protocol: HTTP
VPC: web-server-vpc
Protocol version: HTTP1
Health check: HTTP
Click **Next**

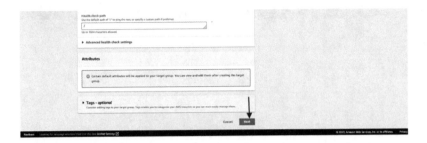

The **Register targets** page appears. Select all three EC2s and click **Create target group.**

Repeat creating a target group with HTTPS by clicking on **Create a Target** on the **Target groups** page. Add the target with the name EC2-TargetGroup-2. The **Target Groups** page now show both target groups created.

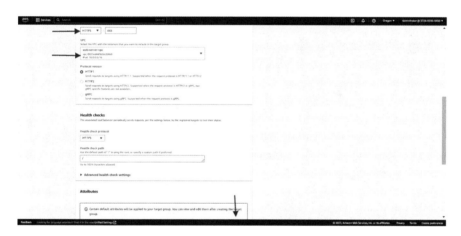

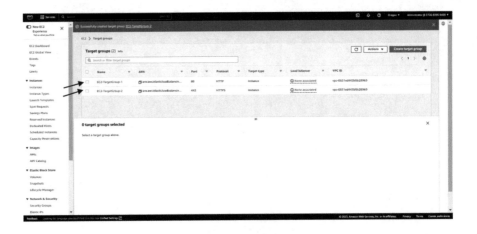

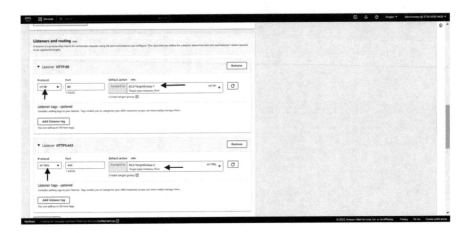

Incoming web traffic will now be load balance across all available EC2s.

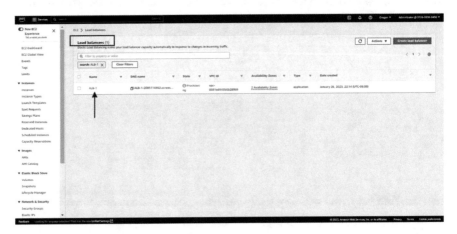

Task 8: Create and Configure AWS Aurora Database

AWS has many choices of databases. You will use the Relational Database Service (RDS) with Aurora. Aurora is proprietary to AWS but is many times faster and cheaper than commercially available databases such as Oracle.

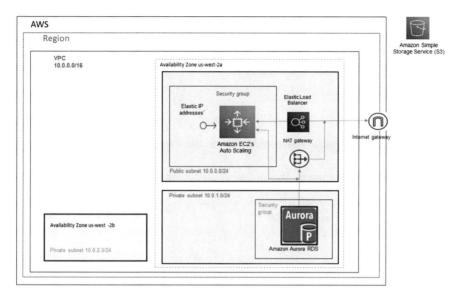

In the navigation menu, under **Amazon RDS**, choose **Subnet groups**. Click on **Create DB Subnet Group.**

The **Create DB subnet group** page appears. Under **Subnet group details**, provide the following information:

Name: rds_sg_1
Description: RDS subnet group
VPC: web_server_vpc

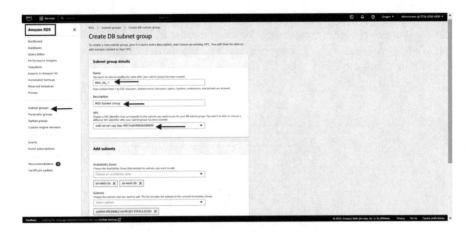

Scroll down to **Add subnets.** Include the following information:

Availability Zones: us-west-2a and us-west-2b
Subnets: 10.0.1.0/24 and 10.0.2.0/24
Click Create

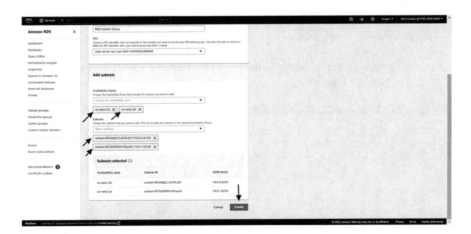

Note: To create an RDS database in the same VPC requires two available subnets, each in a different availability zone. RDS is deployed in the 10.0.1.0/24 subnet in **us-west-2a** AZ. In a single AZ deployment subnet, 10.0.2.0/24 in the **us-west-2b** AZ does absolutely nothing for RDS. The reason for the second subnet so that should you decide to migrate a redundant multi-AZ RDS deployment, the subnet is available to do so with no disruption to the infrastructure.

In the navigation menu, click on **Databases.** The **Subnet groups** page will show that **rds_sg_1** has been created.

Step 2: Create a MySQL database instance.
On the **Databases** page, click on **Create database.**

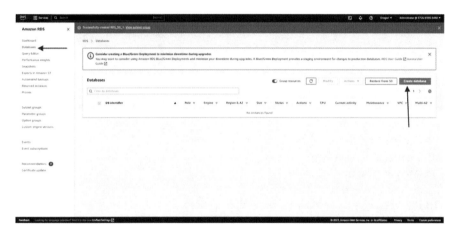

On the **Create database** page, under **Choose a database creation method**, select "Standard create." Under **Engine options**, choose "Amazon Aurora."

Scroll down this page and provide the following information:

Edition: Amazon Aurora MySQL-Compatible Edition
Capacity type: Provisioned
Available version: Aurora (MySQL 5.7) 2.07.2
Templates: Choose Free Tier if available, otherwise choose Dev/Test
Database cluster identifier: database-1

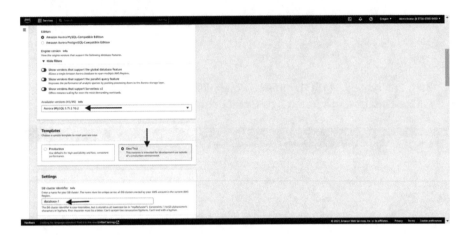

Master username: admin
Auto generate password: No
Master password: AWS12345
Confirm password: AWS12345
DB instance class: Memory optimized, db.r5.large

Availability & durability: Don't create an Aurora replicate.
Virtual private cloud: web-server-vpc
Subnet group: rds-sg-1

Public access: No
VPC security group: Choose existing
Existing VPC security groups: default
Availability zone: us-west-2a (this is where Aurora will be deployed)

Database authentication: enable Password authentication
Click on **Create database.** This process may take several minutes. Refreshing the
  page is necessary to view progress and completion.

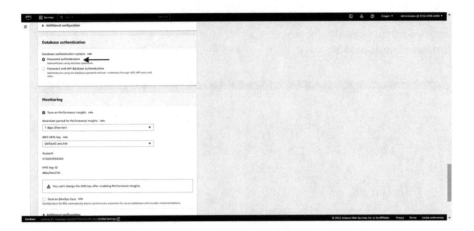

The next page shows that the database has been successfully created and is operational.

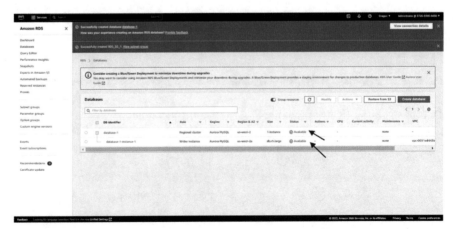

Your Amazon RDS Aurora database has now been deployed.

Task 9: Create and Configure an S3 Storage Bucket and Connect It to EC2

S3 is object storage inside AWS. Individual S3s are referred to as "buckets." S3 is a global service, so it does not reside inside any specific VPC. EC2s in your VPC can access an S3 bucket regardless of where it may be physically located. If permitted, S3 can be accessed by anyone from anywhere on the Internet. Therefore, care must be taken to block unwanted access to your S3s.

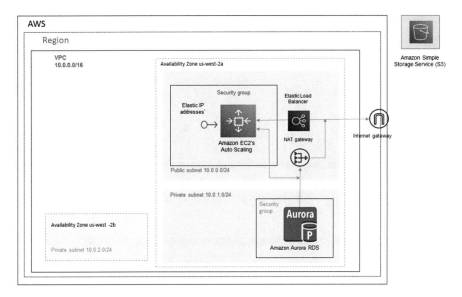

Task 9a: Create and Configure an S3 Storage Bucket

Access S3 from the navigation menu. When the **Buckets** page, opens click on **Create bucket.**

When the **Create Bucket** page appears, choose:
Bucket name: aws-lab-s3. (choose a unique name)
AWS Region: US West (Oregon) us-west-2.
Object ownership: ACL's disabled

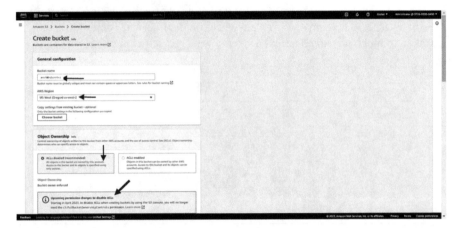

Scroll down on this screen. **Important: Block *all* public access.** Disable **Bucket versioning.**

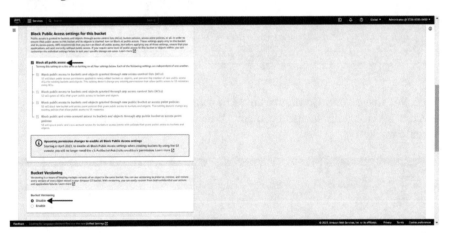

Encryption key type: Amazon S3-manahed keys
Bucket key: Disable
Object Lock: Disable
Click on **Create bucket.**

The next screen confirms that your S3 bucket has been successfully created. Click on the bucket name.

Task 9b: Connect an EC2 to Your S3 Bucket

From the navigation menu, connect to IAM. At the IAM dashboard, click on **Roles.** At the **Roles** page, click on **Create roll.** A role is necessary to allow EC2 permission to access S3.

At the **Select trusted entity** page, select **AWS service**, and then under **Common use cases**, select "EC2" and then click on **Next.**

The **Add permissions** page appears. Search for, find, and select **AmazonS3ReadOnlyAccess**. Click **Next.**

The **Name, review, and create** page appears. For **Role name**, enter "S3ReadOnlyAccess." Review the other entries. Scroll to the bottom of the page and click on **Create roll.**

The **Roles** page appears and validates the role that has been successfully created.

Task 9c: Connect and EC2 to S3

From the navigation menu, access the EC2 dashboard and choose **Instances.** Select the EC2 instance you want to connect to S3 (we will connect to webserverEC2_1). Click on **Actions,** then **Security,** and then on **Modify IAM role.**

The **Modify IAM role** page appears. From the **Choose IAM role** drop-down menu, select "S3ReadOnlyAccess." Click on **Save.**

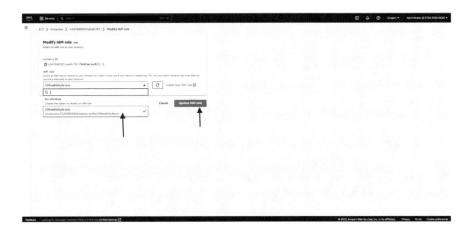

The **Instances** page appears and confirms that S3ReadOnlyAccess has been attached to the webserverEC2_1 instance. Check the box next to **webserverEC2_1** and then on **Security.** See that permission has been applied. **webserverEC2_1** now has access to the S3 you created.

The creation and configuration of all the components of a three-tier web architecture on AWS is now complete.

Clean Up

When you complete the lab, you should go back and delete everything you created. Otherwise, charges against either your free or your pay-as-you-go account will accumulate. It can be a shock to discover you have an AWS bill for hundreds of dollars. Follow the below steps:

You will delete what you created, basically in reverse order. Delete services in the following order:

Task 1: S3
Task 2: RDS Aurora
Task 3: Application load balancer
Task 4: Auto Scaling EC2 instances
Task 5: NAT gateway
Task 6: Internet gateway
Task 7: Security groups
Task 8: Subnets
Task 9: VPC and route tables
Task 10: Elastic IP addresses
Task 11: Verify that all of the above have been accomplished.
Task 12: Closing the AWS account

Task 1: Delete S3 Storage

Access S3 from the navigation menu, click on **Buckets,** and check the box next to bucket name **aws-lab-student-s3** and click on **Delete.**

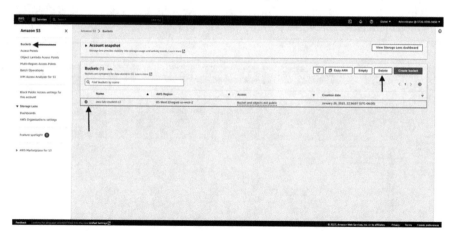

The delete confirmation page will appear.

Type the bucket name **aws-lab-student-s3** in the text box. When the **Delete bucket** highlights, click on it.

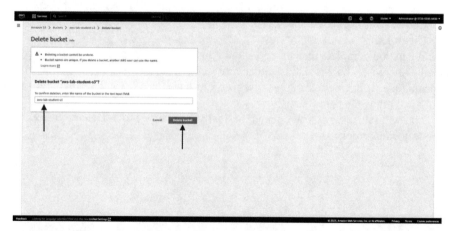

The **Bucket** page will appear confirming the bucket has been deleted.

Task 2: Delete the RDS Aurora Database

From the navigation menu, go to **Amazon RDS** and then **Databases**. Select **database-1-instance-1**. Under **Actions**, select **Delete.**

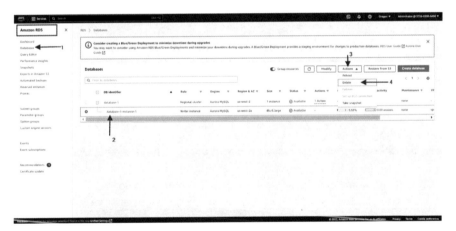

The delete confirmation page will appear. Do not create a final snapshot. Select the acknowledgement state and type "delete me'" in the text box. Click on **Delete.**

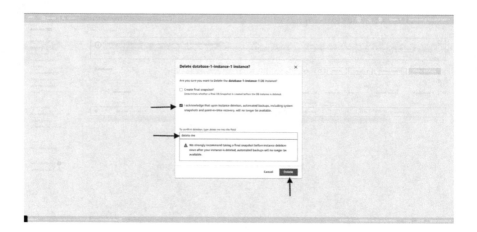

The Databases page will appear, showing that the database deletion is in progress.

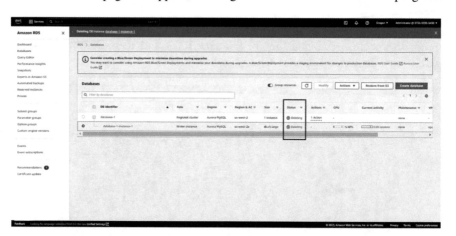

It will take some time, wait for it and below prompt will appear.

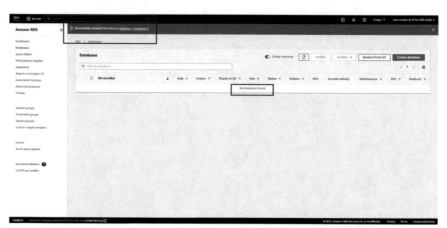

Now click on **Subnet groups** in the navigation menu. Select the **rds_sg_1** subnet group and click on **Delete.**

The delete subgroup page appears. Click on **Delete.**

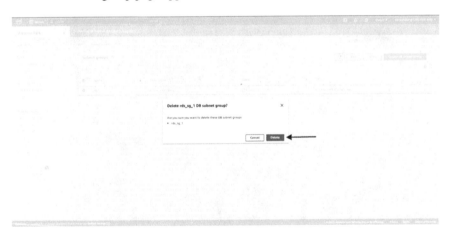

The **Subnet group** page appears. Verify that the subnet group has been deleted.

Task 3: Delete the Application Load Balancer

Find **Load balancers** on EC2 on the navigation menu. Select the load balancer. Under **Actions**, select **Delete**.

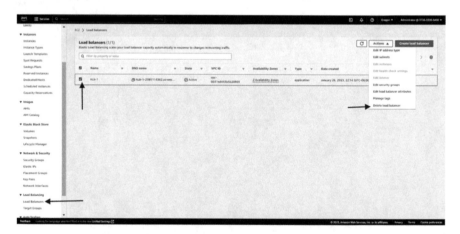

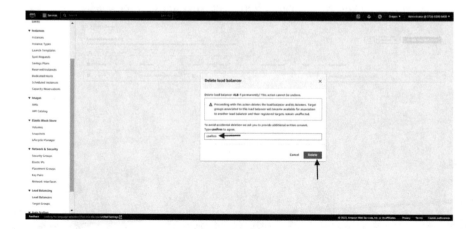

Under **EC2 Load balancer** on the navigation menu, click on **Target groups.** Select both EC2-TargetGroup-1 and EC2-TargetGroup-2, then **Delete.**

The delete verification page appears. Click on **Yes, delete.**

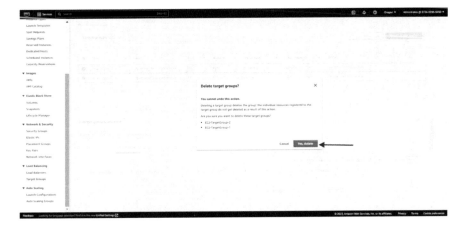

The **Target groups** page will appear. Verify that both target groups were deleted.

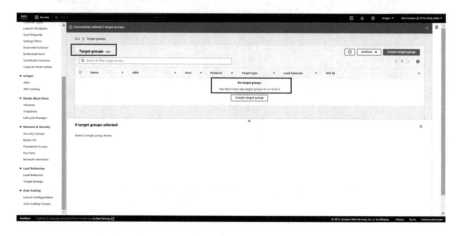

Task 4: Delete Auto Scaling EC2 Instances

Go to the EC2 on the navigation menu and select **Auto Scaling groups.** Check the box beside **ASG1**, then **Delete.**

The **Delete Auto Scaling group** warning appears. In the text field, type "delete" and then click on **Delete.** The **Auto Scaling groups** page will show that three EC2 instances are being deleted. This will take several minutes to complete, but when finished, no EC2 instances should be present.

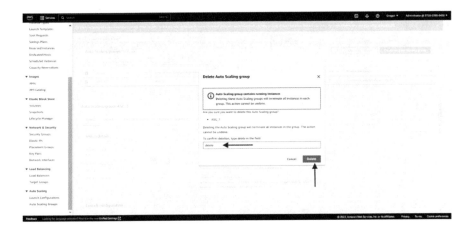

From the navigation menu, open the **Launch configurations** under **Auto Scaling.** Check the box beside any existing launch configurations. Under **Actions**, click on "Delete launch configurations."

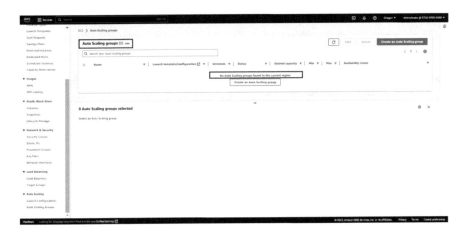

The **Delete launch configurations** warning appears. Click on **Delete.**

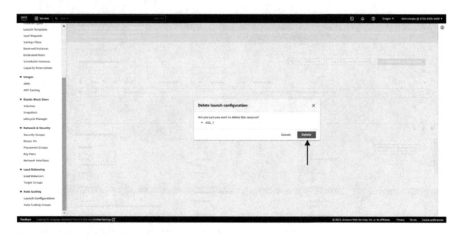

The **Launch configurations** page will appear. It should not show any existing launch configurations.

Task 5: Delete the NAT Gateway

In the navigation menu, find the VPC dashboard and then go to **NAT Gateways**. Select **NAT_1**, and under **Actions**, click "Delete NAT gateway."

The Delete NAT gateway warning appears. In the text box, type "delete" and then click on **Delete.**

After a few minutes, verify that NAT_1 has been deleted.

Task 6: Delete the Internet Gateway
From the VPC dashboard, click on **Internet Gateways**. Click the button next to **IGW_1**. Under **Actions**, select **Detach from VPC.**

The **Detach from VPC** warning appears. Click on **Detach Internet Gateway**.

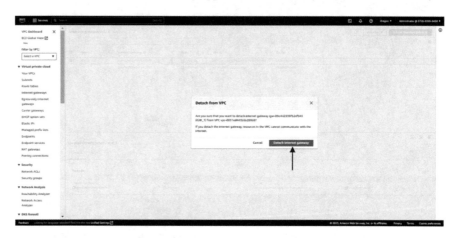

The **Internet Gateway** page appears. Select **IGW_1**. Under **Actions**, click on "Delete internet gateway."

The **Delete internet gateway** warning will appear. In the text box, type "delete" and the click on **Delete internet gateway.**

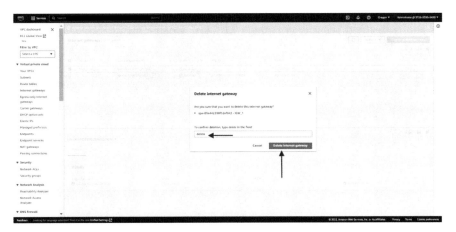

The **Internet gateways** page will appear reflecting the Internet gateway has been deleted and no others are present.

Task 7: Delete Security Groups

From the navigation menu go to **Security** and then **Security Groups.** The default security group belongs to the default VPC and cannot be deleted.

Notice that both security groups are "default." Default security groups cannot be deleted.

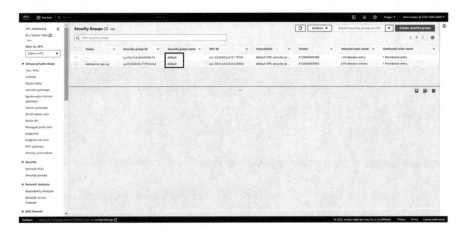

Task 8: Delete Subnets

From the VPC dashboard, click on **Subnets**. Click on each subnet entry check box. Under **Actions**, click on **Delete subnets.**

A delete warning page appears. Enter "delete" in the text box and click on **Delete**.

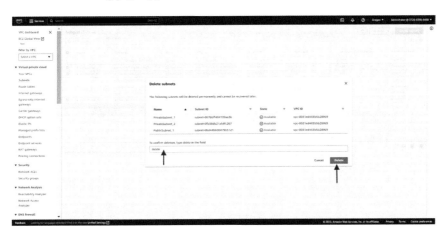

The **Subnets** page appears and verifies that the subnets have been deleted.

Task 9: Delete the VPC and Related Route Tables
From the VPC dashboard, click on **Your VPCs**. The default VPC can be but should not be deleted. Select the web-server-vpc, and the under **Actions**, click on **Delete VPC**.

The **Delete VPC** page appears warning the related route tables will also be deleted. Type "delete" in the text box and click **Delete**.

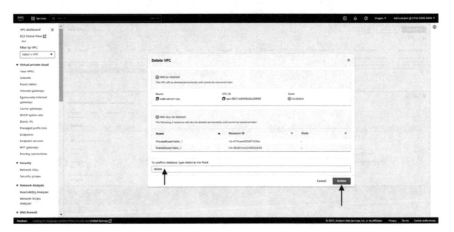

Under the VPC dashboard, click on **Route Tables.** All route tables have been deleted except for the default VPC default route table.

Under **Security** under **Virtual Private Cloud** on the navigation menu, select **Security Groups**. Notice that the default security group for the VPC that was previously undeletable is now deleted. Only the default VPC security group remains.

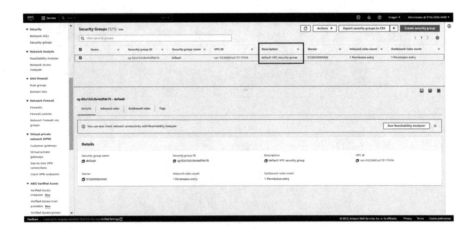

Task 10: Release Unneeded Elastic IP Addresses

Unused elastic IP addresses will incur charges. From the **Virtual Private Cloud** on the navigation menu, click on **Elastic IPs.** Release elastic Ips. Click on one, and under **Actions**, click on **Release Elastic IP address.**

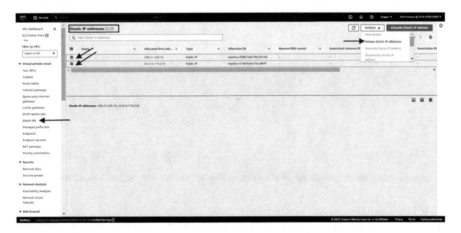

The **Release elastic IP addresses** warning appears. Click **Release**.

The **Elastic IP addresses** page appears and verifies that there are no elastic IP addresses left.

Task 11: Review Your AWS Account Services.

Navigate to the VPC dashboard. If you have created any services for your study purposes, delete them. Other than that, all that is left are default services. Unnecessary charges can now be avoided.

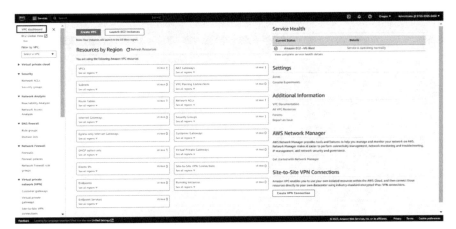

Task 11: Closing Your AWS Account

To close your AWS account

- Open the Billing and Cost Management console at https://console.aws.amazon. com/billing/home#/.
- On the navigation bar in the upper-right corner, choose your account name (or alias), and then choose Account.

- Scroll to the end of the page to the Close Account section. Read and ensure that you understand the text next to the check box. After you close an AWS account, you can no longer use it to access AWS services.
- Select the check box to accept the terms and choose Close Account.
- In the confirmation box, choose Close Account.

9.3 Summary

The lab began with directions on how to create an AWS Free Tier account. Upon completing this lab, you will have created an AWS free tier account, which was followed by an overview of the AWS Management Console. In addition, AWS Identity and Access Management (IAM) was used to create user accounts and apply security.

AWS was used to create and configure a three-tier website. You began by creating a virtual private cloud (VPC) to which were added compute virtual machines with automatic web servers that would be created when the EC2 is turned on. Load balancing and autoscaling of EC2s was added. Security of the VPC as well as EC2 services was implemented with security groups. An RDS Aurora database was added, as well as S3 object storage that is accessible from the EC2s. Finally, to avoid unexpected and unnecessary AWS charges, all work was deleted.

Chapter 10
Microsoft Azure Lab

10.1 Azure Free Account

To begin, you will need to set up an Azure account. New users have $200 in credit to spend for the first month and access to many products thereafter for 1 year.

To sign up, you will need a phone number, a credit card for identity verification, and a Microsoft or GitHub account. Your credit card won't be charged unless after your free trial period you elect to continue using Azure on a pay-as-you-go basis.

Begin by going to https://azure.microsoft.com/free.

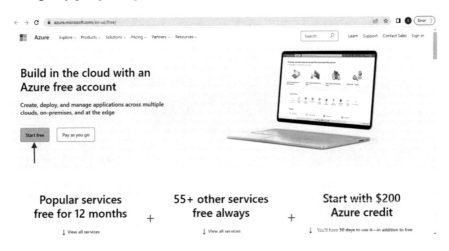

Note: Use a different account ID than your SMU student ID.

© The Author(s), under exclusive license to Springer Nature Switzerland AG 2024
M. S. Kingsley, *Cloud Technologies and Services*, Textbooks in
Telecommunication Engineering, https://doi.org/10.1007/978-3-031-33669-0_10

1. Click **Start free**.
2. Sign in with your Microsoft or GitHub account, or create a free Microsoft account.
3. Provide the requested information on the **out you** page. Click **Next** to continue.
4. On the **Identity verification by phone** screen, provide a telephone number and select whether you want a text or callback to receive a verification code.
5. Provide the code in the **Verification code** box and click on "Verify code."
6. When verified, enter your credit card information and click on **Next**.
7. Now accept the **Agreement** and privacy statement. Click on **Sign up**.

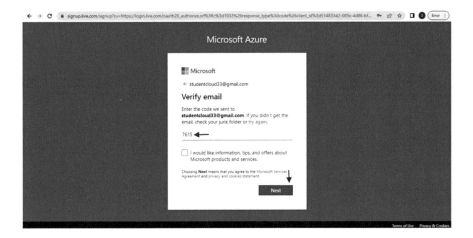

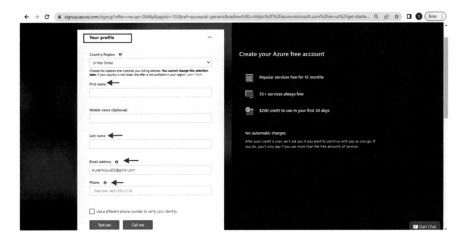

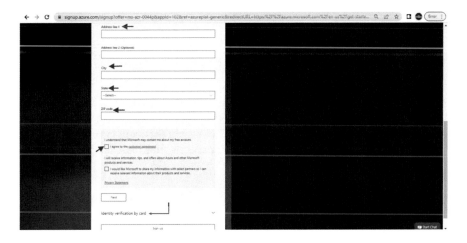

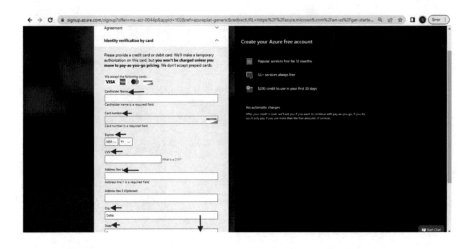

Congratulations! You have successfully set up a free account and should be on the Azure portal home page.

10.2   Azure Portal Overview

Access the Azure portal homepage by going to https://portal.azure.com/. A page similar to the one below will appear. Rapid access to previous operations as well as **Navigate** and **Tools** is easily accessible. Clicking on the **icon**(1) in the upper left corner will present a drop-down menu with access to all resources in the portal.

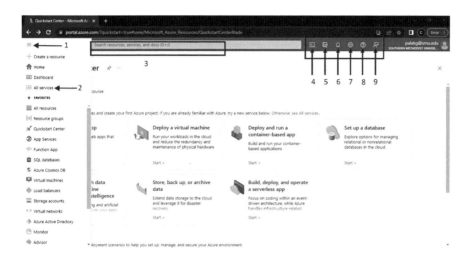

Clicking **All services(2)** will present a page with the most recently used services at the top, followed by all resources by category. Rapid access to resources by category can be accomplished using the menu on the left of the page.

Rapid access to any service can be accomplished by entering the name in the **search bar**(3). Several icons at the top right side of the page are available.

The **>_ icon** (4) on the left will access the Command Line Interface (CLI).

Directories and subscriptions(5) are accessed by clicking on the funnel icon.

The **bell icon**(6) will show any notifications such as system health issues detected.

The **gear icon**(7) accesses system settings including portal appearance, language, and other information.

The **question mark icon**(8) provides customer support and troubleshooting information.

Clicking on the **person icon**(9) will drop down a form to provide feedback to Microsoft.

10.3   Azure IAM

In this lab, we will work on Azure Active Directory (AD) and role-based access control (RBAC) in Microsoft Azure.

Navigate to the Azure Service – **Azure Active Directory**.

To create users, click on **Users**.

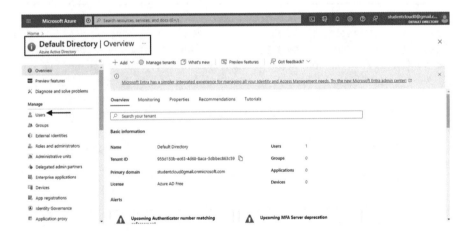

After the User page appears, click on **+New User** and then **Create new user**.

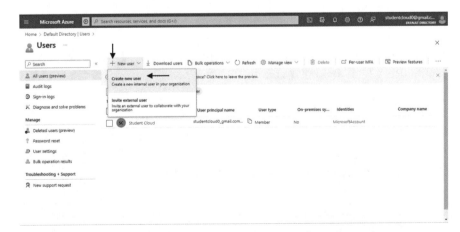

Provide the Username as **Azure1**, keep the default domain as it is.

Also, provide the Name of the user as **Azure1**.

Click on **Let me create the password** and specify the Initial Password as per your choice.

Disable Block Sign In. Click on Create User.

User **Azure 1** has been created.

Now, click on **Groups** under Azure Active Directory.

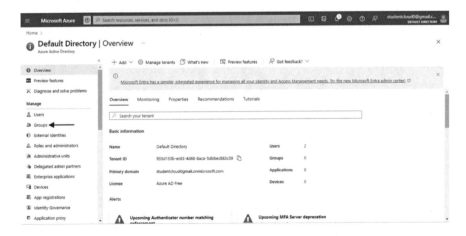

Click on **+ New Group**.

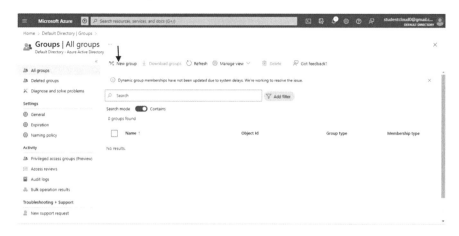

When the New Group page appears, fill in the following information:

Group Type: **Security**
Group name: **Storage Account – Support**
Group Description: **(your choice)**
Under Members, add the User **Azure 1** and select.

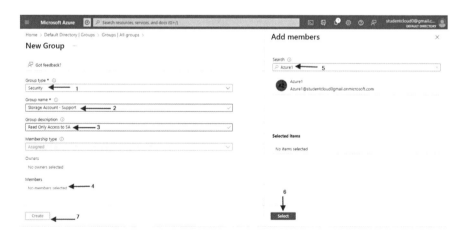

You have created a group.

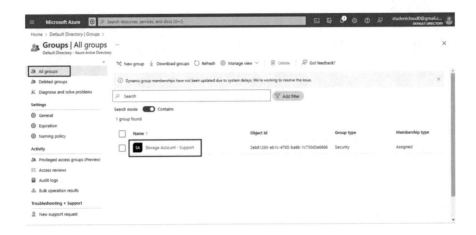

Navigate to Subscription in the search bar and click on **Subscription**. Select the available subscription, here **Azure subscription 1**.

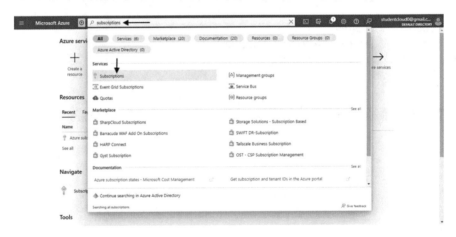

When the page appears, click on **Access control (IAM)** on the left-hand panel.
When the Access control (IAM) page appears, click on **Add** and then **Add Role Assignment**.

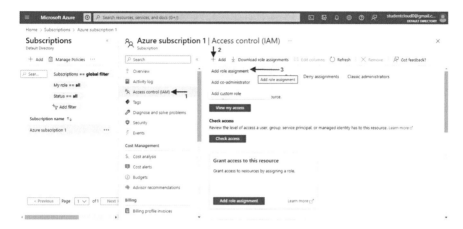

On Add role assignment page, select Role as **Owner** and select **Next.**

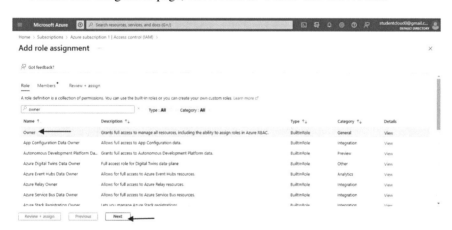

Assign access to **User, Group, or Service Principal.**
Finally, select the **+Select members**, the right-side window will open, select for **Storage Account – Support**, which we created earlier, and click on **select** and then **Next**.

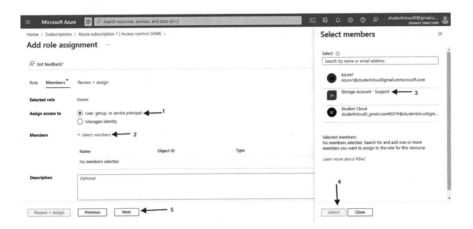

A review page will appear; review the information, and click on **Review + assign**.

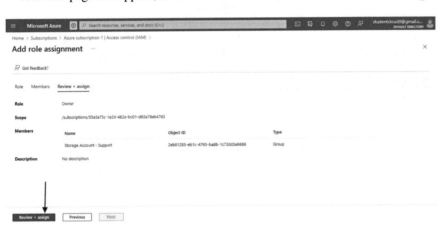

Under Azure Active Directory, click on **Users**.
Under Users, click on user "Azure 1".

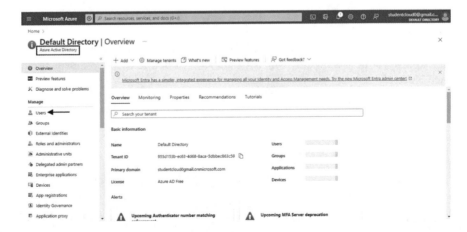

Get the fully qualified domain name (FQDN) for the **Azure1** user. Login with the "Azure 1" user.

Put the Azure 1 username and password. Update Azure 1 password.

Verify that you're logged in as an "Azure 1" user.

10.4   Microsoft Azure: Monster Lab

10.4.1   Lab Introduction

Similar to the generic three-tier website previously demonstrated, the Microsoft Azure Lab three-tier website architecture is illustrated in the figure below:

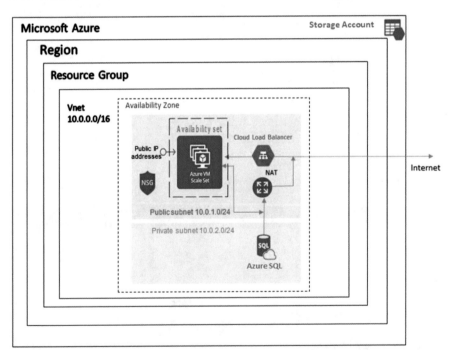

You will implement the basic Azure architecture. Crate and configure the following:

- **Task 1: Resource group**
- **Task 2: Virtual Private Cloud (VPC) network**
- **Task 3: VPC security rules**
- **Task 4: NAT gateway**
- **Task 5: Compute VMs as Scale Sets**
- **Task 6: Web server instances**
- **Task 7: Load balancer**
- **Task 8: Instance group with autoscaling**
- **Task 9: SQL database instance**
- **Task 10: Cloud Storage bucket**

Task 1: Create a Resource Group

Azure has a defined cloud management hierarchy (see figure below). When your organization first acquires an Azure account, root access is assigned, which has unlimited administrative privileges. Root access should never be used. Instead, access should be granted to separate management groups using IAM. Management groups might be Human Resources, IT, Finance, and so on. Each management group will then have different subscriptions. Subscriptions are used to separate administrative factors of each management group department, such as operational costs. Next on the management hierarchy is the resource group. At this level, an IT administrator can create logical groups or resources such as VMs, storage volumes, IP addresses, and so on. Resource groups make it easier to apply access controls, monitor activity, and track costs related to workloads. Resource groups allow easier and more effective assignment and management of role-based access controls (RBACs) for users.

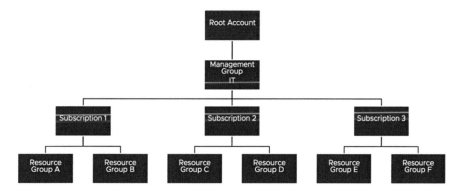

10.5   Azure Management Hierarchy

A resource group resides in an Azure region, as illustrated below:

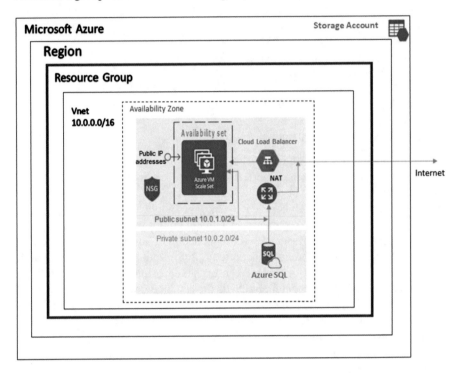

Login to the Azure Portal with your credentials. Click on **All services**. In the search area, begin entering "Resource." Click on **Resource Groups**.

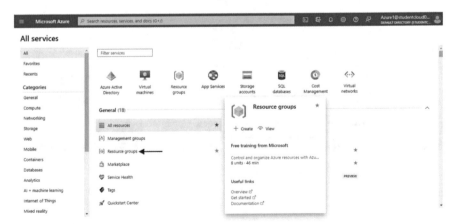

When the "Resource Groups" screen appears, click on **+ Create.**

On the "Create a resource group" screen:

Subscription: Azure subscription 1
Resource **group:** ResourceGroup1
Region**:** East US
Click **Next: Tags >**

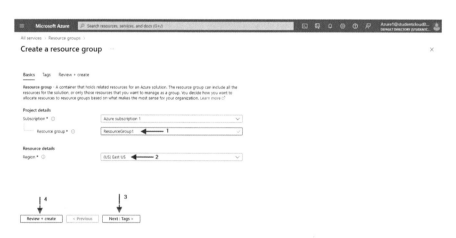

Tags are used to provide tracking capabilities to your resource group. For now, leave it blank and click **Next: Review + create** >. When the next screen appears, click **Create.**

The next screen will reflect the resource group that has been created in the **East US** region.

Return to the Azure **Home** screen.

Task 2: Create a Virtual Network (VNet)

The VNet is your logical slice of the Azure infrastructure. Most of your Azure services will reside in the VNet.

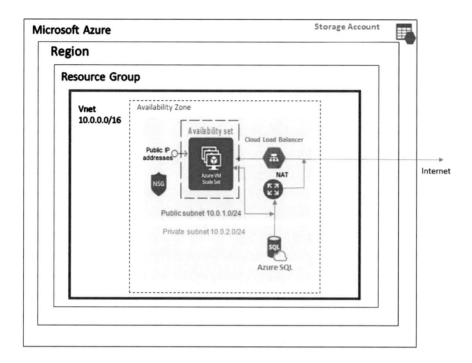

Click on **Virtual networks**. If the icon is not visible, type "virtual" in the search area to find **Virtual networks**.

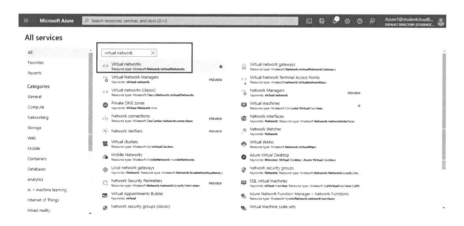

On the screen "Virtual networks," click on **+Create**.

When the screen "Create virtual network" screen:

Project details:

> **Subscription**: Azure subscription 1
> **Resource group:** ResourceGroup1

Instance details:

> **Name:** Vnet1
> **Region:** East US

> Click **Next: IP Addresses >**

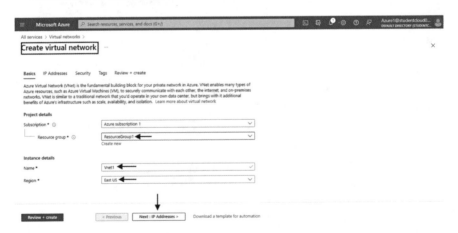

Subnets text

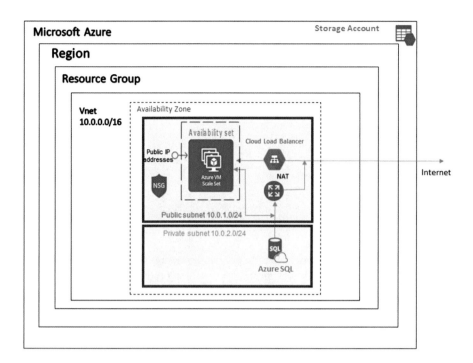

Notice on the next screen that the **IP address space** is "10.0.0.0/16." Next, we will add subnets. Click on **+ Add subnet**.

A small screen on the right side will appear:

Subnet name: Public-subnet
Subnet **address range:** 10.0.1.0/24
Click **Add.**

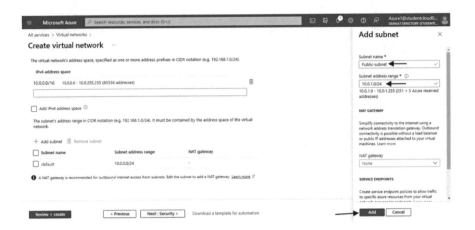

On the refreshed screen, notice that the public subnet has been added to the VNet. Now click again on **Add subnet**. On the small screen on the right:

Subnet name: Private-subnet
Subnet address range: 10.0.2.0/24
Click **Add.**

You should see the private subnet added to Vnet1. Click on the **Next: Security** tab on the bottom.

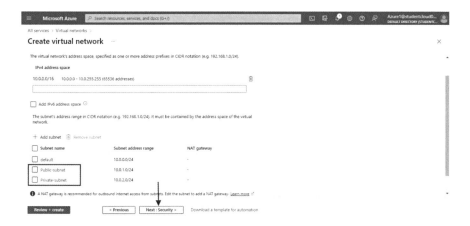

On the screen, disable all security. Click **Tags**, and remain all the fields blank.
Click **Review + create.**

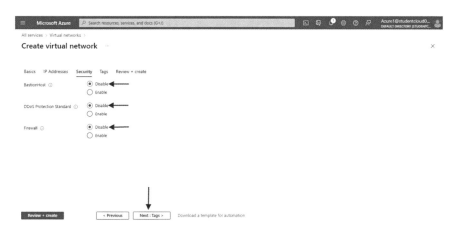

On the review screen, verify that the settings have been captured correctly and
click **Create**. After a few minutes, you should get the confirmation below that Vnet1
was successfully created. Under Next steps, click on "Go to resource."

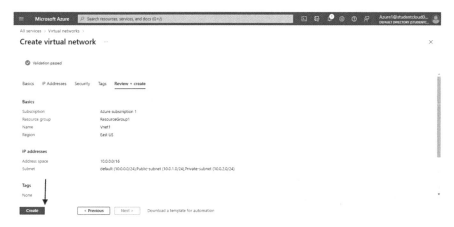

The deployment process will take place, and after some minutes, it will be completed.

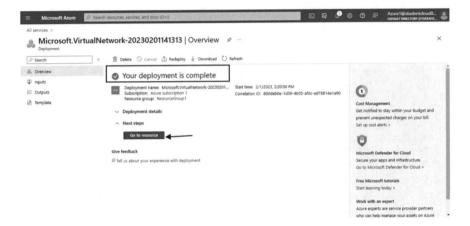

Notice **Vnet1** is connected to "ResourceGroup1." In the left panel, click on **Subnets**.

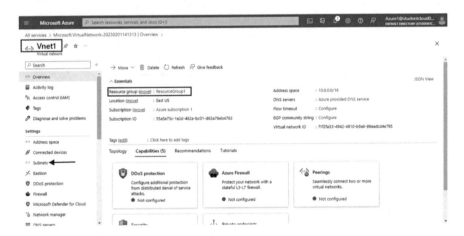

Both the public and private subnets should be present.

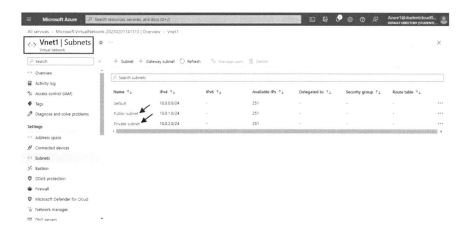

Task 3: Create Security Group

Next, begin typing "Network security" in the search field at the top of the screen.

From the pop-up, click on **Network security groups**.

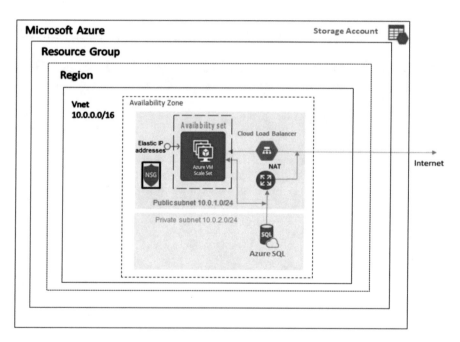

From the Network Security Groups menu, click on **+ Create**.

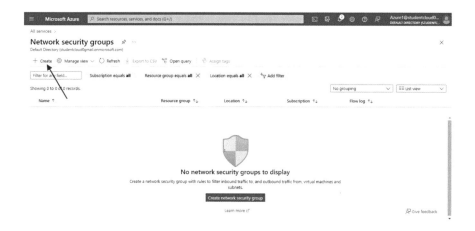

When the **Create network security group** menu appears:

Project details:

> **Subscription**: Azure subscription 1
> **Resource group:** ResourceGroup1

Instance details:

> **Name:** NSG_Public
> **Region:** East US

Click **Review + create**.

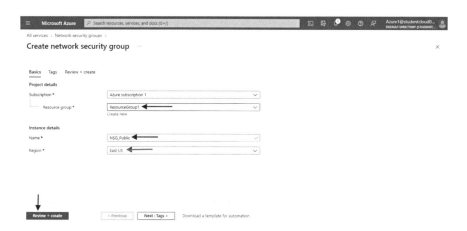

Verify the network security group is validated successfully and click on **Create**.

After a short delay, the network security group should deploy successfully. Click on **Go to resource**.

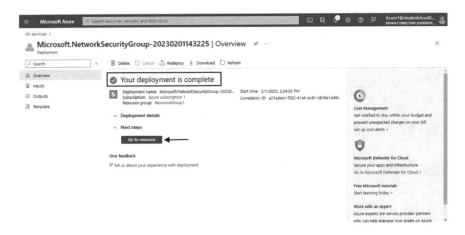

Now that you have created a network security group around your public subnet, the inbound and outbound traffic rules need to be implemented.

Notice that the default **Inbound security rules** for the public subnet are to allow all traffic from inside Vnet1 to Vnet1 (traffic coming into a subnet from another subnet) and traffic coming from a load balancer. All other traffic, including incoming traffic to the VNet from the Internet, will be blocked. Therefore, an inbound rule needs to be added to allow traffic coming from the Internet to access the VNet.

The default Outbound Security Rules allow outbound traffic inside the VNet (traffic leaving a subnet destined for another subnet) as well as traffic leaving the VNet bound for the Internet.

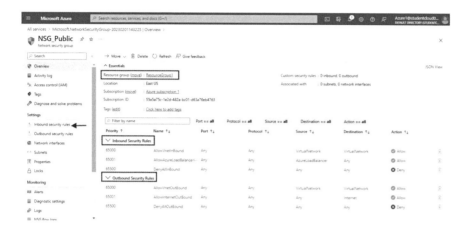

Click on **Inbound security rules** in the left panel and then **+ Add**. A side screen will pop up. Add the following inbound HTTP port 80 rule to the network security group to allow inbound Internet traffic to reach Vnet1:

Source: Any
Source port ranges: *
Destination: Service Tag
Destination service tag: VirtualNetwork
Service: HTTP
Destination port ranges: 80
Action: Allow
Priority: 100
Name: Port_80
Clicl **Add**

Verify that HTTP (port 80) has been added to the **Inbound security rules**.

HTTPS also needs to be enabled. Click again on **+ Add** under **Inbound security rules**. When the side screen pops up, enter the following information as was done for HTTP:

Source: Any
Source port ranges: *
Destination: Service Tag
Destination service tag: VirtualNetwork
Service: HTTPS
Destination port ranges: 443
Action: Allow
Priority: 120
Name: Port_443

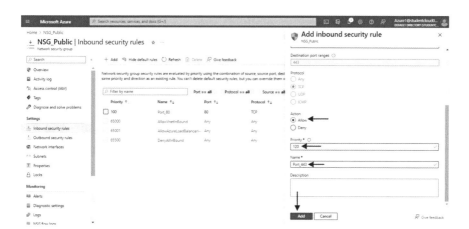

Verify that the HTTPS (Port 443) inbound security rule has been added.

Note: The Azure network security group destination was "Virtual Network." Unlike similar features of other cloud providers that can only be applied at the subnet level, the Azure network security group can be applied at the VNet, virtual machine, or subnet level.

Task 4: Create and Configure a NAT Gateway

Network Address Translation (NAT) allows access to the Internet from private subnets. However, NAT is installed on a public subnet. We will need NAT later.

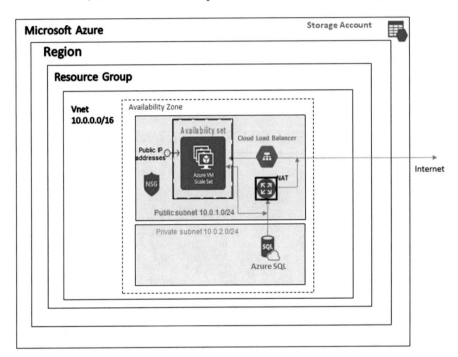

Begin typing "NAT" in the search field at the top of the page. When the pop-up appears, click on **NAT gateways.**

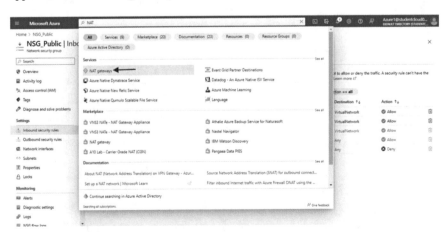

When the **NAT gateways** screen appears, ensure the **Subscription** is correct and click **Create.**

When the "Create network address translation (NAT) gateway" screen appears:

Subscription: Azure subscription 1
Resource group: ResourceGroup1
NAT gateway name: NAT-Gateway
Region: East US
Availability zone: No Zone
Idle timeout (minutes): 10
Click on **Next**: **Outbound IP** at the bottom of the page.

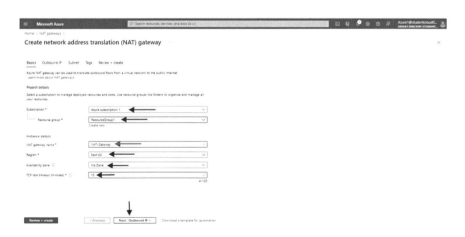

When the next screen appears:
   Under **Public IP address**, click on **Create a new public IP address:**

Name: NAT-PublicIP

Click **ok**.
Click **Next: Subnet** at the bottom of the page.

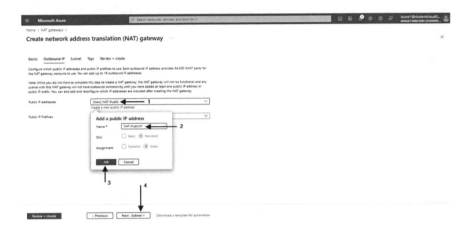

When the next page appears:

Virtual networks: Vnet1
Enable **public subnet**.
Click on **Review + create.**

Review the NAT gateway configuration and verify the validation passed. Click on **Create**.

After a delay of a few minutes, verify that the NAT gateway was successfully deployed. Click on **Go to resource group**.

When the "ResourceGroup1" page appears, verify that **NAT-Gateway**and **NA-PublicIP** are present.

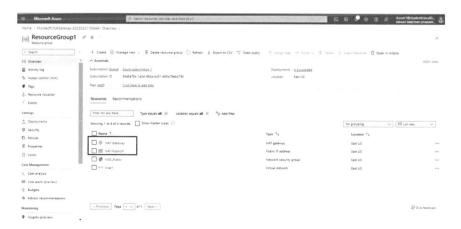

Task 5: Create and Configure Compute Virtual Machines as Scale Set
Website software will be hosted on Azure compute virtual machines. Preparations will be made so load balancing and autoscaling can be implemented when needed. Therefore, Azure Scale Sets containing two compute VMs will initially be provisioned in a few more steps later.

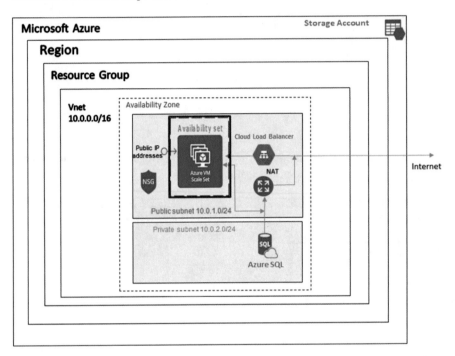

From the Azure portal, begin entering "Scale" in the search area at the top of the screen. Click on **Virtual machines scale sets** when it appears.

When the "Virtual machines scale sets" screen appears, click on **+ Create**.

This will allow you to create a virtual machines scale set. A scale set is a group of VMs that can be load balanced or autoscaled later.

Enter the following information:

Under "Basics":

Subscription: Azure subscription 1

Resource group: ResourceGroup1

Virtual machine scale set name: WebScaleSet

Region: (US) East US

Availability zone: None

Orchestration mode: Uniform

Security type: Standard

Image: Windows Server 2022 Datacenter – Gen 2 (Select free images under "see all images")

VM architecture: x64

Run with Azure Spot discount: Do not select

Size: Standard_B1s – 1vcpus, 1 GiB memory (free service eligible)

Username: Azure

Password: "Your choice"

Licensing: Do not select

At the bottom of the screen, select **Next | Disks >** .

The next screens reflect these inputs.

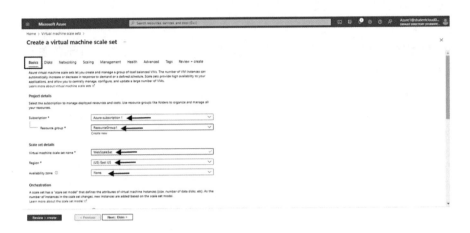

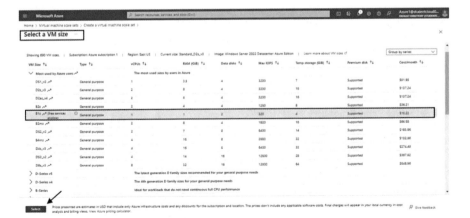

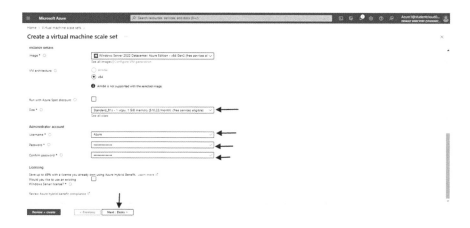

When the screen changes:

OS disk type: Premium SSD (locally redundant storage)
Key management: Platform-managed key
Click **Next: Networking >**

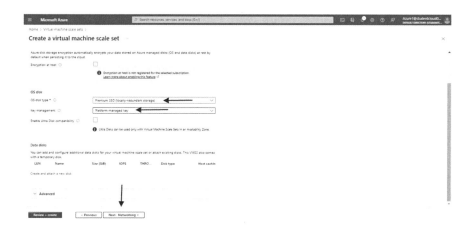

Under Networking:

Virtual network: Vnet1
Under **Network interface**, click the **Edit** symbol for "Vnet1-nic01."

The **Edit network interface** screen should appear. Configure the following:

Name: Vnet1-nic01
Virtual network: (default present) Vnet1
Subnet: (default) 10.0.0.0/24
NIC network security group: Advanced
Configure network security group: NSG_Public
Public IP address: enabled
Click **OK**

The **Edit network interface** screen should appear. Configure the following:

Name: Vnet1-nic01
Virtual network: (default present) Vnet1
Subnet: (default) 10.0.0.0/24
NIC network security group: Basic

Public inbound ports: Allow selected ports
Select inbound ports: RDP(3389)
Public IP address: enabled
Click **OK**

When the screen reappears, scroll down to the **Load Balancing** section. Enable **Azure load balancer** under **Load balancing options**.

New side-screen will appear:

Load balancer name: WebsScaleSet-lb
Keep other settings as they are and click on **create**.

Click on **Next: Scaling >** at the bottom of the screen.

Under **Scaling**:

Initial instance count: 1

Scaling policy: manual
Scale-in policy: Default – Balance across availability zones and fault domains, then delete the VM with the highest instance ID.
Click on **Next: Management > .**

When the screen repaints provide the following data to **Management** fields:

Enable basic plan for free: enabled
Upgrade mode: Manual- Existing instances must be manually upgraded
Boot diagnostics: Enable with managed storage account (recommended)
Any further fields: Do not enable
Click: **Next: Health.**

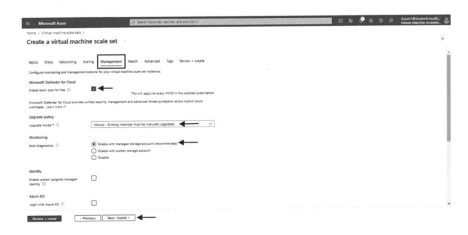

The screen refreshes, Do nothing here. Click on **Next: Advanced**.

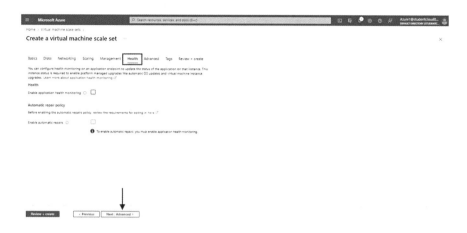

The screen will refresh. Under Advanced fields, enter:

Spreading algorithm: Fixed spreading (not recommended with zones)
Fault domain count: 5
The **Custom data** field is left blank. Ignore any other fields on this page. Click
   Review + create.

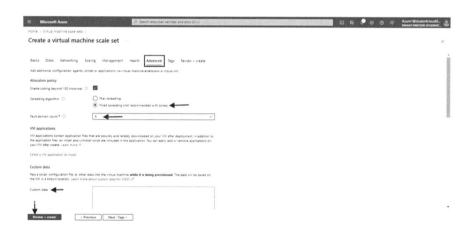

Review next screen, after the validation passes, click **Create.**

Azure will begin checking your configuration. This will take several minutes to complete. Once complete, Azure will show that the set deployment was successful.

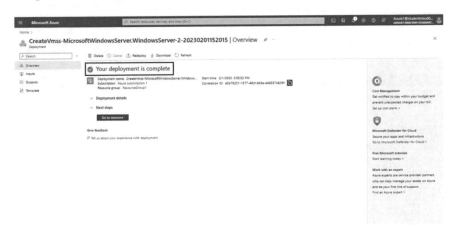

Task 6: Launch Web Server Instances
Now you will launch your web server VM instances.

Under **Recent resources**, click on "WebScaleSet." Otherwise, begin typing "WebScaleSet" in the search field until it pops up and then select. Click on **Instances.**

You should see the compute web server VMs created here: **WebScaleSet_0** and it is running. Click on **WebScaleSet_0**

When the screen refreshes, you will see detailed information about the WebScaleSet_0 VM and various performance monitoring charts for its operation. Click on **Connect** and select **RDP.** RDP is Microsoft's Remote Desktop Protocol that will allow you to connect to your web servers.

When the new screen appears, click on **Download RDP file** and then open the file.

A remote desktop screen message will appear. Click on **Connect**.

When connected, you will see a login screen. Enter the username **Azure1** and the password you previously set up. If you get a certificate error, click "Yes".

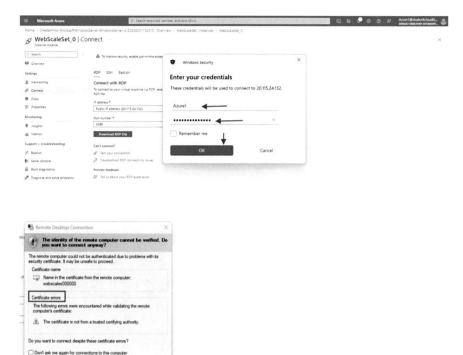

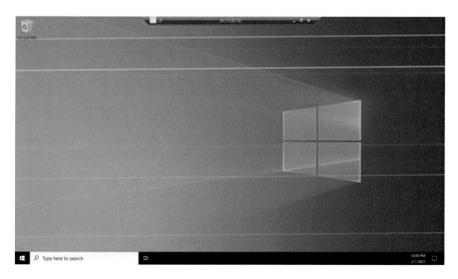

In a few seconds, you will see a Windows screen. If it does not load automatically, search for it to get the Windows Server Manager screen. Click on **Add roles and features**.

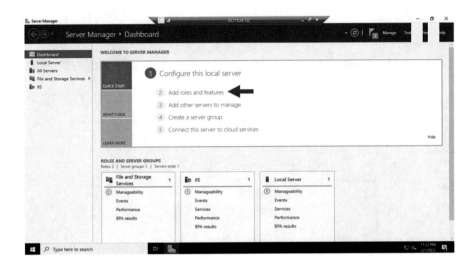

When the "Before you begin" screen appears, click **Next**.

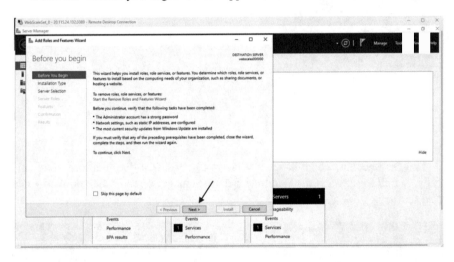

On the "Select installation type" screen, select **Role-based or feature-based installation.** Click **Next**.

On the "Select destination server" screen, choose **Select a server from the server pool**, then select **webscales**, which you created earlier. Click **Next**.

On the "Select server roles" screen, choose **Web Server (IIS).** Internet Information Service (IIS) is Microsoft's web server application. When a second screen appears, select **Add features** and then **Next**.

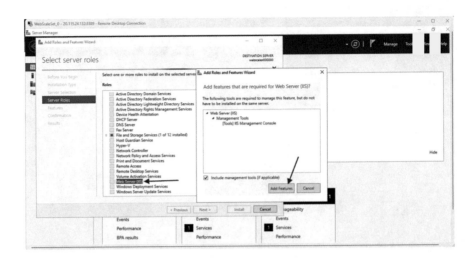

When the "Select features" screen appears, choose those as shown below, then click **Next.**

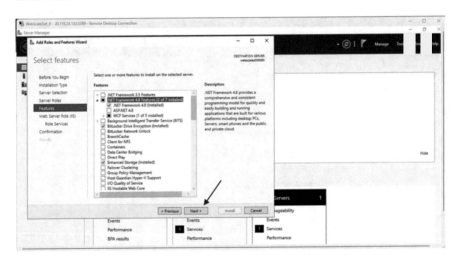

The "Web Server Role (IIS)" screen appears. Read and click **Next.**

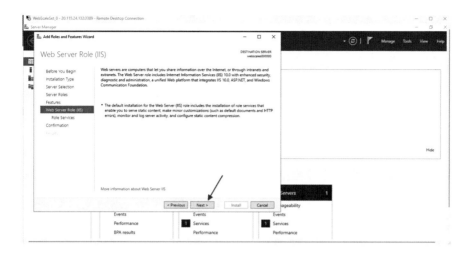

When "Select Role Services" screen appears, leave the options as configured and then click **Next**.

When the "Confirm installation selections" screen appears, click **Install**. The IIS web server will begin to download. The installation progress will be viewable. Success will be presented below the progress bar. When completed, click **Close**.

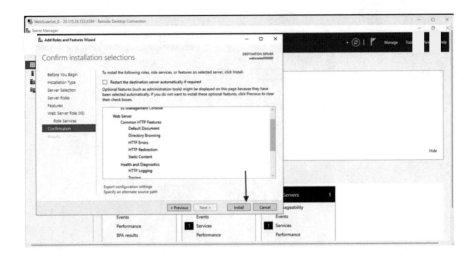

A Windows IIS web server has been successfully installed on compute VM WebScale_0.

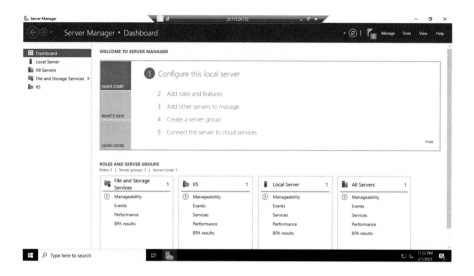

Task 7: Create and Configure a Load Balancer

A load balancer will share the traffic request load between the two servers. If more web servers are added, such as when autoscaling is enabled later, the traffic load will balance across all web servers.

The figure below illustrates the logical location of the load balancer.

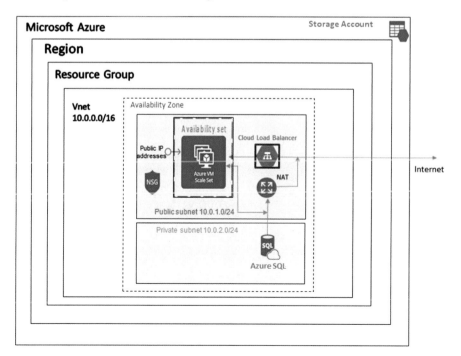

At the Azure portal **Home** screen, begin typing "Load" in the search area. When another screen pops up, select **Load balancers**.

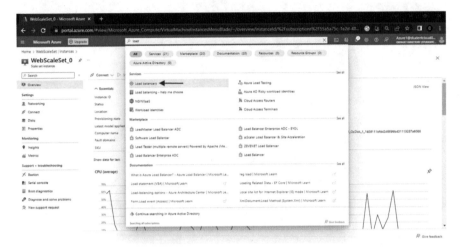

On the "Load balancer | Load balancing" screen, you will see the load balancer (**WebScaleSet-lb**) created previously in the VM Scale Sets task. Click on **+Create.**

When the screen repaints, configure the following fields:

Subscription: Azure subscription 1
Resource group: ResourceGroup1
Name: Public_1_lb
Region: East US
SKU: Basic
Type: Public
Tier: Regional
At the bottom of the screen click **Next: Frontend IP Configuration.**

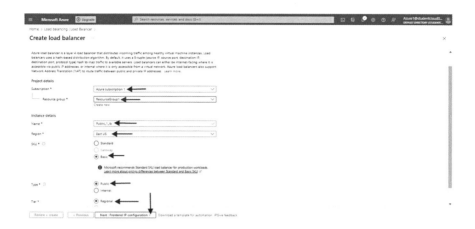

On the "Create load balancer" under "Frontend IP configuration" screen, click **+ Add a frontend IP configuration**. When **Add frontend IP address** screen pops up on the right side, enter the data in or enable the following fields:

Name: Public_1_Load_Balancer
IP Version: IPv4
Public IP address: Create new > Public_1 > ok
Click **Add**

When the screen refreshes, click **Review + create,**

When the screen refreshes, it should reflect that Azure has validated the configuration. Click **Create.** Azure should return after a few minutes with "Your deployment is complete."

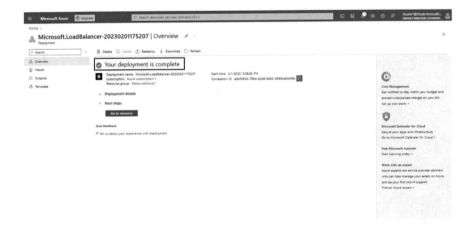

To view your load balancer configuration details go to the Azure portal **Home**, then go to **Load balancers.** Click on **Public_1** load balancer. On the next screen, click on **Frontend IP configuration**. Notice the public IP address of the load balancer. This is the IP address that others can use to reach your web servers via the Internet. The traffic will then be load balanced across both web server VMs.

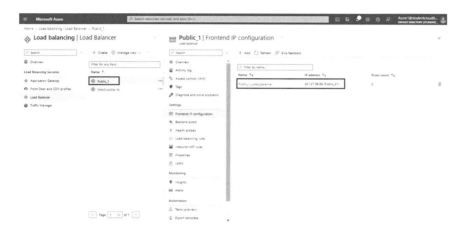

Next, click on **Backend pools** and click on **Add**. The "backend" is the private subnets of the VPC.

When the **Public_1 | Backend pools** screen appears:
The **Add backend pools** screen will appear. Enter the following information:

Name: BackendPool_1
Virtual network: Vnet1 (ResourceGroup1)
Click **Add.**

Backend Pool Configuration: NIC
IP Version: IPv4
Leave all other fields blank.
Click **Add**

   Azure will begin validating your configuration. When validated, you will see that
BackendPool_1 has been has been added to BackendPools.

Task 8: Create and Configure an Instance Group with Autoscaling
On the Azure portal **Home** screen, begin typing "autoscale" in the search area. Click
Autoscale when it appears.

The "Monitor | Autoscale" screen will appear. Click on **WebScaleSet**.

Switch from "Manual Scale" to "Custom autoscale." Enter the following data on the **Autoscale setting** screen in the following fields:

Autoscale setting name: Leave as default (WebScaleSet-Autoscale-350)
Resource group: ResourceGroup1
Scale mode: Scale based on a metric
Instance limits: Leave all fields as "2"
Click on "**Add a rule**" (This will determine when compute VMs will be added or deleted with increases and decreases in traffic.)

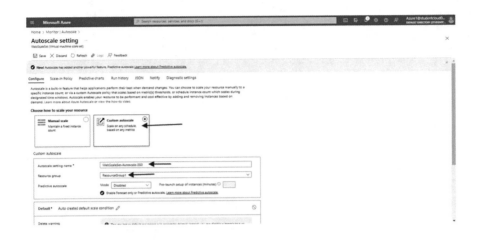

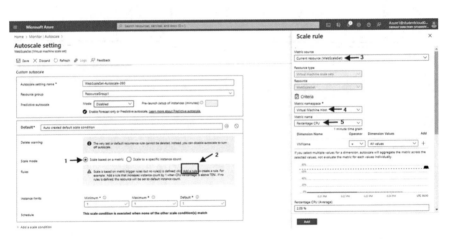

The **Scale rule** screen will appear on the right side. Ensure the following fields are configured as follows: (Parameters that could not be changed are not included.)

Metric source: Current resource (WebScaleSet)
Metric Namespace: Virtual Host Machine
Metric name: Percentage CPU
Operator: greater than
Metric threshold to trigger scale action: 70
Duration: 5
Time grain statistic: Average
Time aggregation: average
Operation: Increase count by
Cool down (minutes): 5
Instance count: 1
Click **Add**

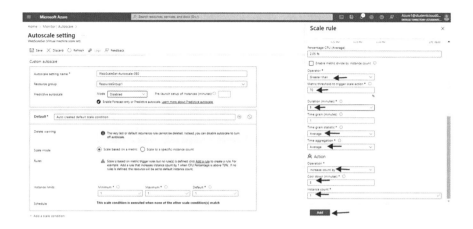

When the screen refreshes, notice that for **Rules** the scale out rule "(Average) Percentage CPU > 70" has been added. Change the **Instance Limits-Maximum** to "5."

This is a scale-out rule. When the average CPU usage of a compute VMs exceeds 70% for more than 5 min, Azure will enable one more web server VM. Azure will continue to add compute VMs as each has a CPU usage that exceeds 70%, up to a maximum of five web server VMs. Scaling-out allows customers to provide rapid service when web traffic increases without having physical hardware operational and standing by, saving a lot of money.

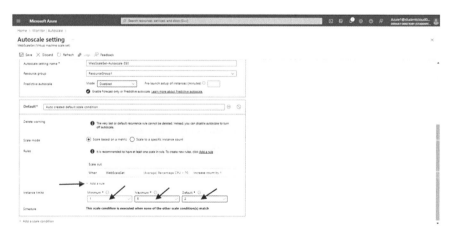

We now need to add a scale-in rule for when web traffic decreases. Click on **Add a rule**.

The **Scale rule** screen appears again on the right side. Configure the fields such that:

Metric source: Current resource (WebScaleSet)
Metric Namespace: Virtual Host Machine
Metric name: Percentage CPU
Operator: less than
Metric threshold to trigger scale action: 70
Duration: 5
Time grain statistic: Average
Time aggregation: Average
Operation: Decrease count to
Cool down (minutes): 5
Instance count: 1
Click **Add**

This is a scale-in rule. When the average CPU usage of a web server VMs decreases below 20% for more than 5 min Azure will remove one web server VM. Azure will continue to remove web server VMs as each has a CPU usage that exceeds 70%, down to a minimum of two compute VMs. Since you only pay for the resources you use, removing VMs allows quickly reducing the cost of resources when demand is low.

When the screen refreshes, notice that the scale-in rule has been added.

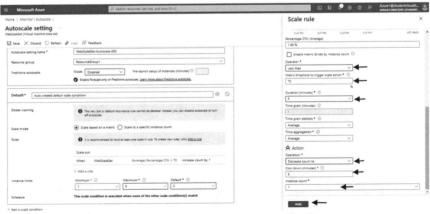

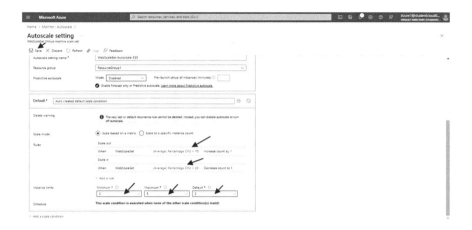

To monitor the usage of your compute VMs scale-in and scale-out activity, go back to the Azure portal **Home** and click on **WebScaleSet**. Click on **Monitoring.**

When the monitoring screen appears, notice that there are a number of metrics monitored. Also notice the time duration of monitoring can be viewed over a few minutes to days.

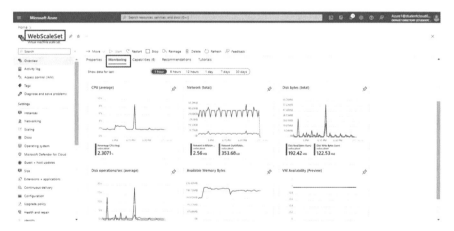

Task 9: Create and Configure an Azure SQL Database

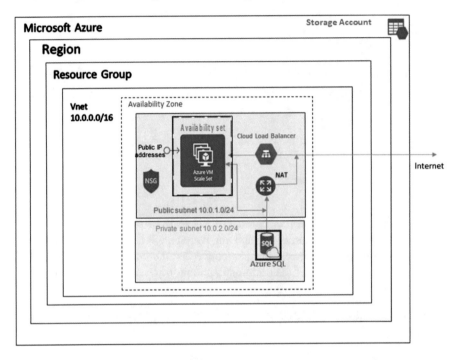

From the Azure **Home** screen, begin typing "SQL" in the search field. Select "SQL databases" when it appears.

On the **SQL Database** screen, click on +Create.

On the "Create SQL Database" screen, provide the following information:

Subscription: Azure subscription 1
Resource group: ResourceGroup1
Database name: SQL_1
Server, select "Create new." When the screen refreshes, enter the following information:

> **Server name**: Server names must be globally unique for all servers in Azure Enter "sql-server-xxxx" and add some random characters like "x" to make the server's name unique. If Azure rejects your server's name, try others until you are successful.
> **Server admin login**: Enter "azureuser."
> **Password**: Enter a password that meets the requirements.
> **Confirm password**: repeat the same password.
> **Want to use SQL elastic pool:** No
> Under **Compute + storage**, click on "Configure database."

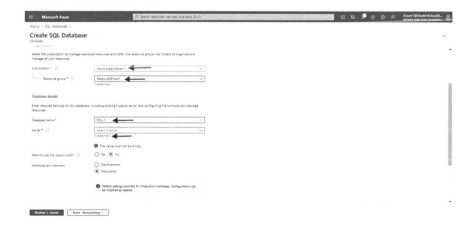

When the screen refreshes, select **Serverless** and click **Apply**. Select **Next: Networking** at the bottom of the page.

When the screen reappears, enter the following information:

Connectivity method: Public endpoint
Firewall rules: Allow Azure services and resources to access this server: No
Add current client IP address: Yes

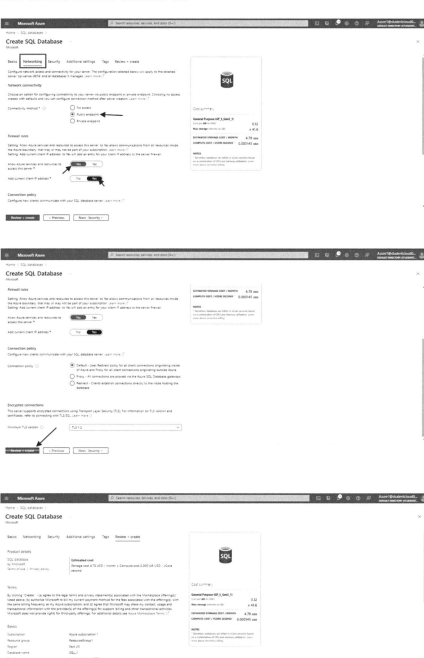

Select **Next: Additional settings** at the bottom of the page.

When the screen refreshes for the **Additional settings** tab, in the **Data source** section, for **Use existing data**, select **Sample**. This creates a sample database, so there are some tables and data to query and experiment with later if desired.

Select **Review + create** at the bottom of the page.

After reviewing, select **Create**. This step may take a few minutes to complete. Once finished, the following screen will appear:

Task 10: Create and Configure an Azure Storage Account

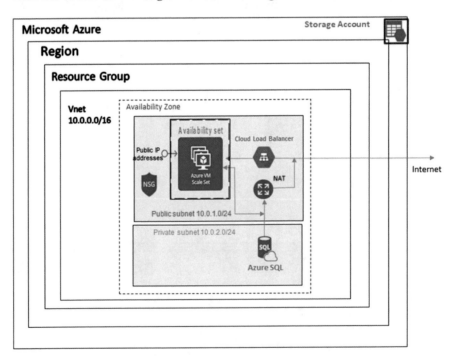

From **Home**, click on **Storage accounts** if the icon is visible or begin typing "storage" in the search field at the top of the page and click on **Storage accounts.**

Ensure the **Subscription** field is correct. Click on **Create** on the upper left or **Create storage account** at the bottom of the page.

When the screen repaints, input the following information:

Subscription: Azure subscription 1
Resource group: ResourceGroup1
Storage account name: Storagexxxxx (Note: Storage account name must be unique globally across Azure. Your storage account name will be different)
Region: East US
Performance: Standard
Redundancy: Locally-redundant storage (LRS)
Click on **Next: Advanced >** at the bottom of the page.

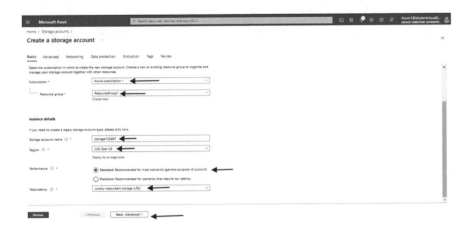

Configure only the following mandatory fields:

Minimum TLS version: Version 1.2
Allow enabling public access on containers: Enable
Access tier: Hot: Frequently accessed data and day-to-day usage
Click on **Next: Networking >** at the bottom on **Networking** a the top of the page.

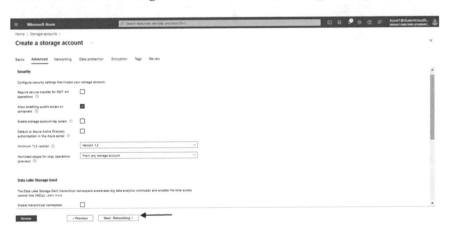

Network Access: Public access from all networks and
Routing preference: Microsoft network routing
Next: Data protection >

Encryption type: Microsoft managed keys(MMK)
Enable support for Microsoft managed keys: Blobs and files only
Click on **Review and then create.**

After several minutes, the storage account will be deployed.

Below will be the result of the storage account that we created. In the next picture below, check out the upload and download sections as well.

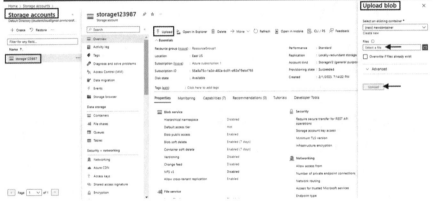

10.6   Deletion

10.6.1   Resource Deletion

Go to the resource groups option from the home page. You can see various resource groups created for different tasks.

Select the resource group created in the first task **ResourceGroup1.** We can see various resources attached to this Resource group.

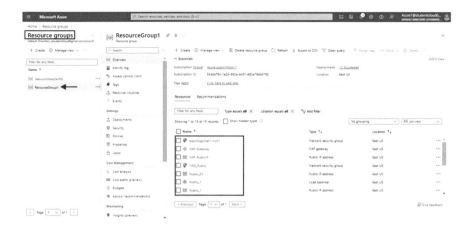

Select **Delete resource group** option at the top. A new tab will open on the right side; type as per the guidance (here **ResourceGroup1**) to confirm the deletion process, and then click on **Delete**.

Under Notifications, you can see that resources are getting deleted one by one. After few minutes, it will be deleted, and you will get the notification.

You can check by going into the **Resource groups** page; there will be no resource group and its related services will be deleted too.

10.6.2   Subscription Deletion

After the subscription is cancelled, billing is stopped immediately. However, it can take up to 10 min for the cancellation to show in the portal. If you cancel in the middle of a billing period, Microsoft will send the final invoice on your typical invoice date after the period ends. After you cancel, your services are disabled. That means your virtual machines are de-allocated, temporary IP addresses are freed, and storage is read-only.

Go to **Subscription**, and select **Azure subscription 1**.

Click on **Cancel subscription**. When the new page appears, select **reason** and click on **Cancel subscription.**

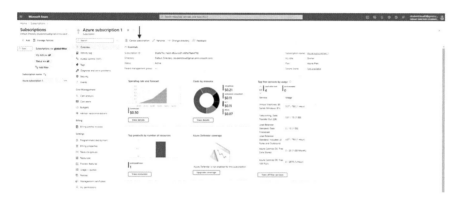

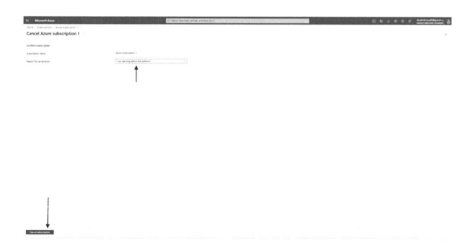

After 10 min, you will see below pictures to confirm your deletion of the subscription.

Chapter 11
Google Cloud Platform (GCP) Lab

By completing this lab you will have experienced the application of the GCP theoretical content previously presented.

11.1 Google Cloud Platform Free Trial Account

To first step in learning the Google Cloud Platform is to create a user account. GCP provides three months or $300 credit of a free trial account access to new users. This should be enough time to learn GCP and decide if you want to continue using it on a pay-as-you-go basis.

Go to cloud.google.com/free and click on **Get started for free**.

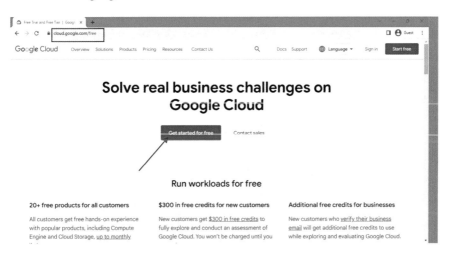

Follow the steps as appeared on the screen:

Step 1 of 2: Sign up for your free Google account using an email account and click **Continue.**

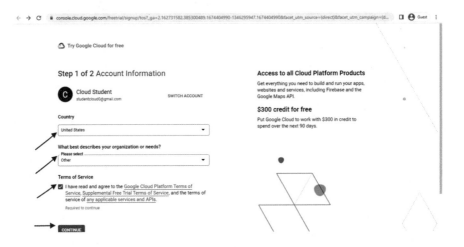

Step 2 of 2: **Payment Information Verification** requires entering a credit card number and other information. Your credit card will NOT be charged unless you explicitly tell GCP that you want to continue on a pay-as-you-go basis when your free trial account expires. Click on **START MY FREE TRIAL.**

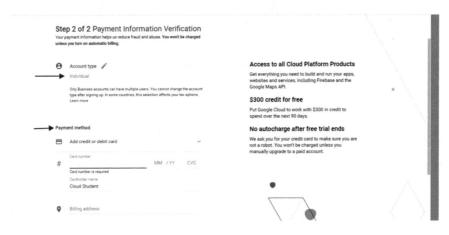

Following snippets will allow to set up basic information:

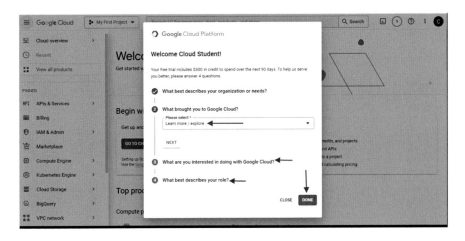

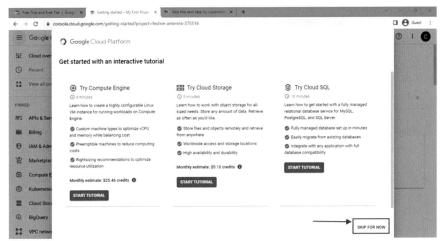

Congratulations! You have successfully created your Google Cloud account. You should see the GCP main menu like the one below.

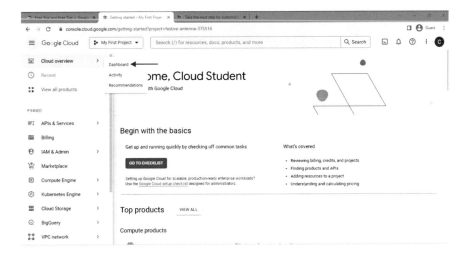

11.2   Google Cloud Platform Console

11.2.1   Overview

Opening the Google Cloud Platform (GCP) console will result in the "Dashboard" screen opening which is a summary of GCP services you have access to.

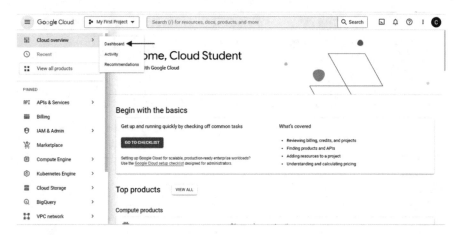

GCP is based on projects. A project can be created, and the results reflected in the **"Project Info"** area. All the services you create will appear under **"Resources"**. Notice the **APIs** overview and the Google **Cloud Platform status** area. Next is the **Monitoring** function that monitors your services.

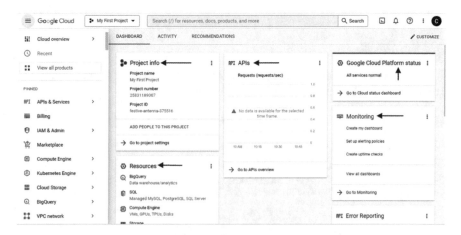

Scroll further down. Cloud **"Trace"** collects latency data from applications and displays it in near real-time. Further down is the **"Getting Started"** section which provides numerous GCP tutorials. "**Error Reporting**" will report problems with your services. **News** relays recent information regarding the Google Cloud Platform; the **"Documentation"** section provides reference resources.

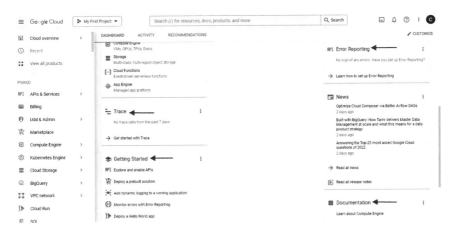

11.3 GCP IAM

GCP operates using the security principle of "least privilege" which states no one should have more access to resources than they need to complete their responsibilities. This goal is accomplished by GCP using Identity and Access Management (IAM) which allows the administrator to enable access to some resources while disallowing access to others.

To understand GCP IAM requires defining several components:

- Resources
- Identities
- Roles
- Policies
- Principals

IAM defines who has access to "resources." Basic resources include compute, storage, database, and networking functions as well as many other ancillary resources. *Who* in IAM is defined as an "identity" while *what* access is allowed is a defined in a "role." Whether the resource can allow requested access is defined in a "policy." IAM access is granted to "principals." A principal can be a user or various groups or accounts in GCP.

Your first task after creating a GCP account is to secure the GCP cloud environment using the GCP Identity and Access Management (IAM) system. Begin by logging into to access the GCP portal and click on the **Navigation Menu** icon in the upper left corner and then on **IAM & Admin**.

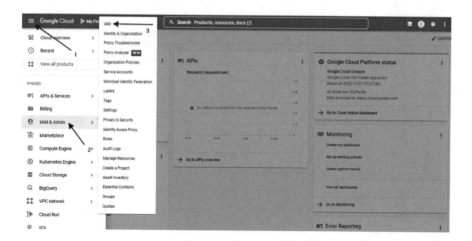

The **Permissions** page will appear. This is where you add users and permission for access to resources.

To create individual user access permissions, click on **"Grant Access"**. Then **Add principles and roles for "My GCP Project"** page will appear. Now enter the email address or GCP provided credentials of the user in the **new principles** field followed by clicking on the **Select a role** field. A few of the most used roles are available.

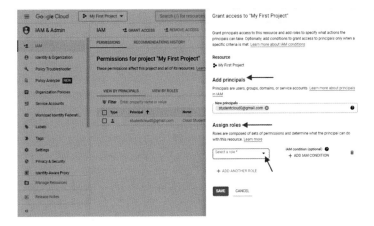

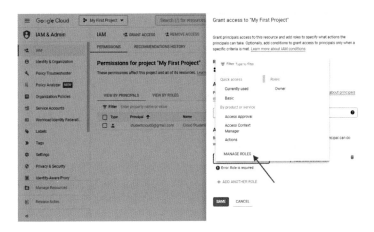

GCP has nearly 1000 other predefined and selectable roles available to assign to users. To view all roles, click on **Manage roles.** The **Roles** for **"My GCP Project"** project page appear.

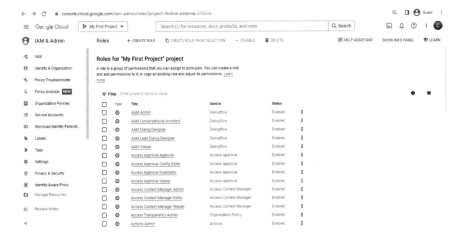

To secure applications from accidental or intentional modification or degradation of applications, **Service Accounts** were created. In contrast to permission and roles for individual users the Service Account is used for access by groups of users such as application developers. Click on **Service Accounts** in the left-hand panel. The **"Service accounts for project "My First Project"** will repaint. Note there may be one or more default service accounts already created for you. Click on **+ Create Service Account**.

The **Create servcie account** screen will appear. Enter an account name In the **Service account name** field such as "MyServiceAccount" used here. GCP will then automatically create a **Service account ID** account designation in the next field. Enter descriptive information for the service account as you choose in the **Service account designation** field. Click on **CREATE AND CONTINUE**.

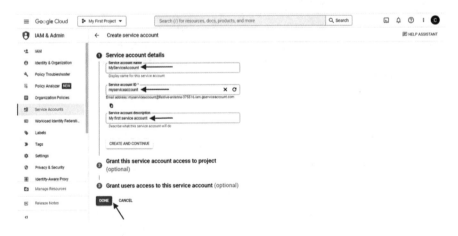

The **Service accounts for project "My First Project"** page will reappear, and the new Service Account is visible. Click on the three vertical dots at the end of the line of the new Service Account that will present a drop-down menu.

Manage details will return to the previous screen. Click on **Permissions.** The **Permission** page will appear to show which principals that can currently access the Service Account. Additional principals can be allowed access by clicking on the **Grant Access** button producing the additional screen on the right. As with users, roles can also be assigned by clicking on the **Roles** button.

Click on **Keys**. The screen will refresh. By clicking on the **ADD KEY** button new or existing encryption keys can be enabled on the Service Account. **Metrics** and **Logs** can be viewed and configured if needed.

From the Navigation Menu click again on **Service Accounts** to verify the account has been created. The Service Account is identified by an email address that application developers with permissions previously created developers can access to deploy their resources.

11.4   GCP LAB

11.4.1   Introduction

The GCP lab is a basic three-tier web architecture similar to the generic model previously discussed. You will implement the basic GCP architecture using the following tasks to create and configure the following GCP resources:

- **Task 1: Create a Project**
- **Task 2: Virtual Private Cloud (VPC) network**
- **Task 3: VPC firewall rules**
- **Task 4: Cloud NAT**
- **Task 5: Compute instance group**
- **Task 6: Web servers**
- **Task 7: Autoscaling**
- **Task 8: Load balancer**
- **Task 9: Cloud SQL database**
- **Task 10: Cloud Storage bucket**

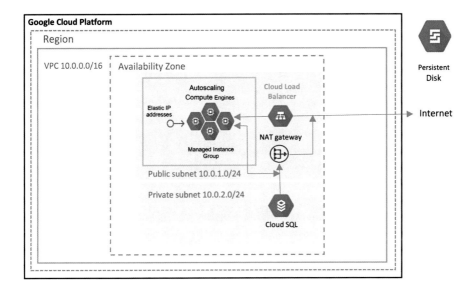

Task 1: Create a Project

Click on the **Navigation menu** icon on the upper left corner, mouse over **Cloud overview,** and select **Dashboard**.

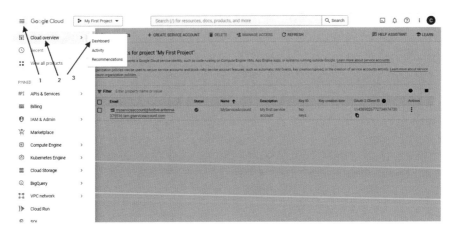

The dashboard page will appear. Click on the **Select a project** drop down menu. Note: A previous project name may appear in the field. In this case it would be **My First Project.** Click on **New Project.**

The **New Project** page will appear. Enter the project name under **Project Name.** **Location** is where the project will be stored. You can choose to save to a **Folder** or an **Organization.**

When the page below appears click **ADD PEOPLE TO THIS PROJECT**. On right hand side, window will pop-up, in the "New principal" field enter the email of who you want to add to the project. The email entered must be associated with a valid Google email. If the user does not have one Google will assign one. In the **Role** field enter "Owner." Click **Save.**

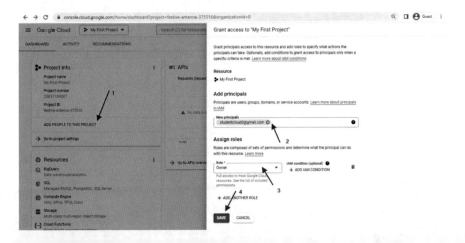

The Dashboard page will appear. Click on the **Select a project** drop down menu to see the **My Second GCP Project"** has been created.

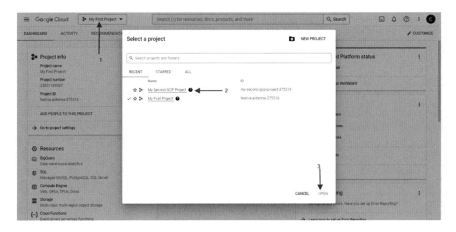

Task 2: Create a GCP VPC Network
Click on "Compute Engine" in the dashboard **Resources pane** on the left of the screen.

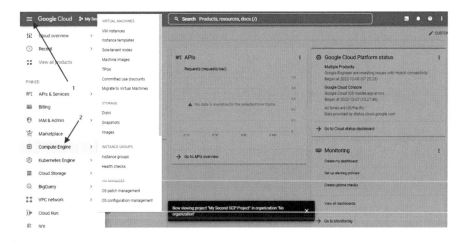

The **Compute Engine API** page will appear. Click **Enable**.

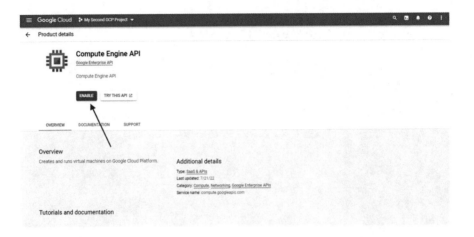

When the screen below appears enter "**vpc networks**" in the Search products and resources field. The **VPC Networks** page appears. Note that GPC will create VPC for your account in every GCP zone. However, we will create your own new VPC. Click on **CREATE VPC NETWORK.**

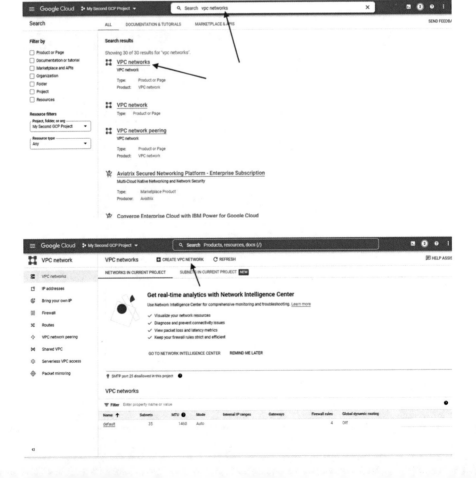

Creating a VPC network includes several steps. First, subnets need to be created. A public subnet will be created to host the web servers and a private subnet that will contain the database.

The IP address space for the VPC is 10.0.0.0/16. A /24 custom subnet CIDR results in the third octet reserved for subnets providing 256 subnets with 256 hosts each. 10.0.1.0/24 to be used for the public subnet and 10.0.2.0/24 for the private subnet. When the screen below appears enter:

Name: my-vpc"
Subnet creation: Custom

Next, under New subnet:

Name: public-subnet
Region: us-central1
IP stack type: IPv4 (single stack)
IP Address range: 10.0.1.0/24
Private Google Access: On
Flow logs: Off
Click **Done**

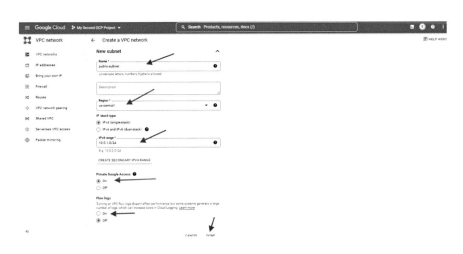

Next, create a private subnet by clicking "**ADD SUBNET**" entering the following information:

Name: private subnet
Region: us-central1
IP stack type: IPv4 (single stack)
IP Address range: 10.0.2.0/24
Private Google Access: On
Flow logs: Off
Click **Done**

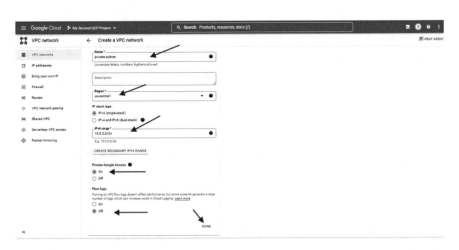

When the screen reappears notice that a private subnet and public subnet has been added to the VPC.

Scroll down until **Dynamic routing mode** is visible and click "**Regional**".

When the **Cloud DNS API** screen appears click **Enable.** This will take a few minutes to complete.

When DNS is enabled, the screen below will appear. Leave **DNS server policy** as "No server policy" and **MTU** as "1460." Click "Create."

After a few minutes the following screen will appear. You will see that "my-vpc" has been created with both public and private subnets.

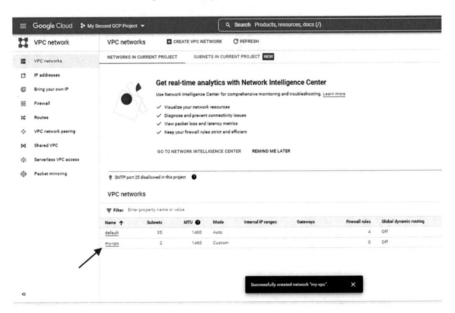

Task 3: Configure VPC Firewalls Rules

The next step is to add firewall rules for the VPC. Traffic leaving (egress) the VPC is *allowed* by default. However, traffic entering (ingress) is *blocked* by default. To service user requests of the web server a rule must be added to the firewall to allow incoming traffic to reach the web servers,

For simplicity, we will add a rule to allow *all* ingress traffic to enter the VPC. In an operational environment, allowing all traffic access to the VPC would not occur. In practice rules would be created to not allow certain protocols or address ranges from accessing the web servers.

In the dashboard left panel click on **Firewall.** As previously noted, GCP creates a default VPC in every zone. The firewall rules that are present apply to them. They allow the VPC subnets to communicate with each other and allow ping and SSH requests. We will create custom firewall rules for our VPC. Click on **Create Firewall Policy**.

When the screen repaints enter the following information:

Policy Name: allow-all
Deployment Scope: Global
Priority: 1000
Logs: Off
Direction of traffic: Ingress
Action on match: Allow
Target Type: All instances in the network
Source filter: IPv4 ranges
Source IPv4 ranges: 0.0.0.0/0 (means "all IPv4 addresses")
Second source filter: None
Protocols and ports: Allow all
Click **Create**.

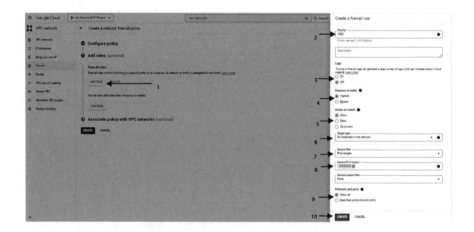

Notice the firewall rule "allow-all" has been successfully created. However, it has not been assigned to anything yet. We will use it later.

Task 4: Implement Cloud NAT

The next step is to add a NAT gateway to allow access from private IP addresses in the private subnet to the Internet using a public IP address. An example of how NAT is used is illustrated in our VPC diagram. The database that we will create is in the private subnet to isolate from outside access. However, it is common that a database would need to access the Internet and then developer to download database upgrades and patches. NAT makes this possible.

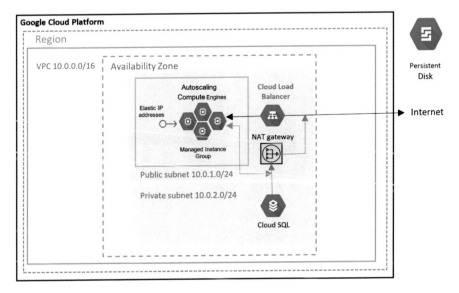

From the **Navigation menu** scroll down to and select **Network services.** When the **Network services** page appears select **Cloud NAT** and select **GET STARTED.**

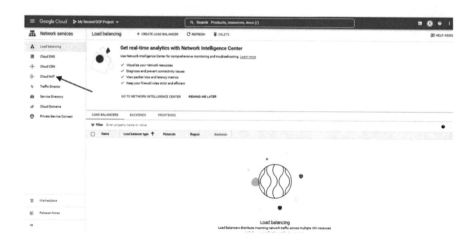

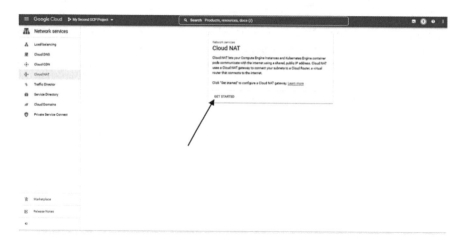

When the **Cloud NAT** page appears enter the following information:

Gateway name: nat-gateway
Select Cloud Router:

> **Network:** my-vpc
> **Region:** us-central1 (Iowa)
> **Cloud router:** Create new router

When a new screen appears enter "router-1" in the Name field and click **CREATE**.
Cloud NAT mapping:

> **Source(internal):** Primary and secondary ranges for all subnets
> **Cloud NAT IP addresses:** Automatic

Destination (external): Internet
Click on **Create**.

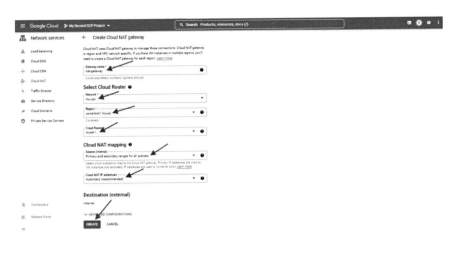

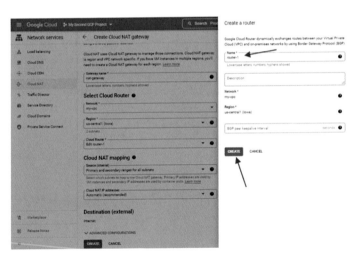

After a few minutes the NAT gateway will be successfully created. Notice the status is "**Running**."

Task 5: Create a Compute Instance Group

You will now create Cloud Engine virtual machines running web servers that reside in your VPC. The first step is to create an instance template. (When the word "instance" is used it refers to a Compute Engine virtual machine.) An instance template allows you to create multiple Compute Engines, in this case with web server applications, quickly instead of creating each one individually. It is similar to AWS AMI's. You will create an instance group of several Compute Engines. However, they will not be used unless the traffic increases and autoscaling will enable them as needed.

Begin typing "instance" in the search area and select **Instance Groups-Compute Engine** from the pop-up window.

When the **Instance Groups** screen below appears choose **CREATE INSTANCE GROUP.**

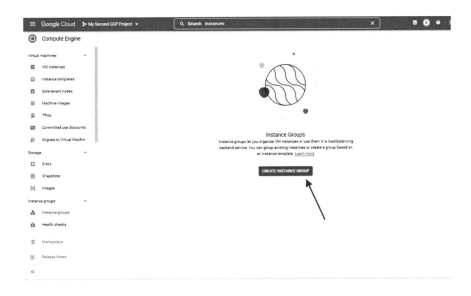

When the **Create Instance Group** page appears, fill the following data:

Name: instance-group-1

From the "**instance template**" drop down menu CREATE A NEW INSTANCE TEMPLATE.

A new half screen will appear. Enter the following information:

Name: Leave as the default "instance-group-1"

Machine configuration:

Machine family: GENERAL PURPOSE

> **Series:** N1
>
> **Machine type:** f1-micro (1 vCPU, 614 MB memory)

Boot disk: click CHANGE. Another half screen will appear.

PUBLIC IMAGES:

> **Operating system:** CentOS
>
> **Version:** CentOS 7
>
> **Boot disk type:** Balanced persistent disk
>
> **Size (GB):** 20
>
> Click **SELECT**

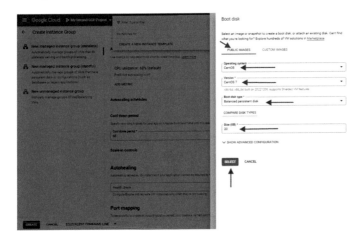

The previous screen again reappears. Notice the boot disk definition has changed.
Continue entering:

Identity and API access/Service accounts/Service account:

> **Service Account:** Compute Engine service account
> **Access scopes:** Allow full access to all Cloud API's

Firewall: Check Allow HTTP traffic and HTTPS traffic

Expand **Advanced options:**
Expand **Networking:**

> **Network Tags:** Blank
> **Hostname:** Blank

IP Forwarding: NOT enabled

Network interfaces:

Click on Edit network interface:
Network: my-vpc
Subnetwork: public-subnet (us-central 1) IPv4(10.0.1.0/24)

Continue to scroll down on this page.

External IP address: Ephemeral
Network Service Type: Premium (If advanced load balancing is selected then you must pay for premium network service.)
Public DNS PTR Record: Enable

Task 6: Create web servers

Scroll down to **Management:**

Expand the **Management** drop down menu.

In the **Automation-startup script** field paste the following script:

sudo -i

yum install -y httpd

systemctl start httpd

systemctl enable httpd --now

echo "Hi from $HOSTNAME" > /var/www/html/index.html

sudo yum install mysql -y

The script will automatically create an Apache web server whenever the template is used to create a Compute Engine.(Note: when copying the script be sure and save is as a plain text Word document and then copy from there.)

Click "SAVE AND CONTINUE" and "CREATE".

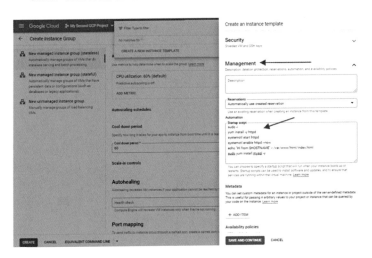

Task 7: Enable Autoscaling

The **Autoscaling** page will appear.

Autoscaling will automatically increase or decrease the number of web servers if traffic reaches a certain threshold of a metric as set by the administrator. In this case the threshold is 60% of CPU utilization.

Enter the following information:

Autoscaling mode: On: add and remove instances to the group.

Minimum number of instances: 1

Maximum number of instances: 4

Autoscaling metrics: CPU utilizations 60% (default)

Continue scrolling down on this screen and enter the information below:

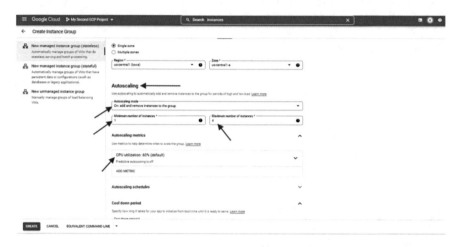

Cool down period: 60
Scale in controls: do not enable.
Under Auto healing, select CREATE A HEALTH CHECk
Health check:

> **Name:** hc1
> **Protocol:** TCP
> **Port:** 80
> **Proxy protocol:** None
> **Logs:** Off

Health Criteria:

> **Check interval:** 10
> **Timeout:** 5
> **Healthy interval:** 3
> **Unhealthy threshold:** 3

Click **SAVE**
Initial delay: 300
Click **SAVE**.

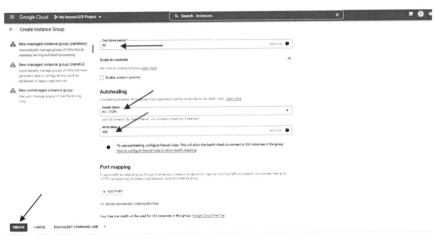

Creating the instance group will take several minutes. When completed you will have created on Compute Engine that can scale and load balance from one to four Compute Engine web server instances. The starting state is one Compute Engine. As traffic increases more Compute Engines will be added as it decreased the number of Compute Engines will also scale down.

When completed the **Instance groups** screen will appear and verify the **instance-group-1** has been created. Refreshing the screen occasionally may be required to see updated status.

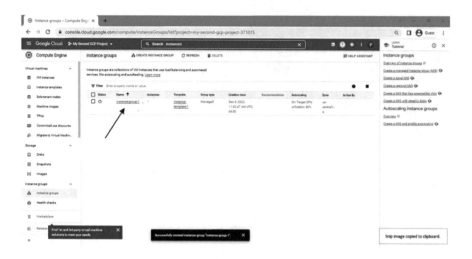

In the left-hand panel click on "VM Instances" to also see the instance group status.

Task 8: Create a Load Balancer

As previously configured, my-vpc initially has only one web server operating. However, as traffic increases autoscaling will enable up to three more web servers, for a total of four, to be added as needed to handle increased traffic. CPU utilization is the metric used. When the web server utilization reaches 60% another web server will be added; when one web server falls below 60% CPU utilization it will be deleted.

As web servers are added to the initial single web server traffic should be load balanced, or divided equally, between the enabled web servers to prevent overloading of some servers while others are underutilized. Therefore, as traffic decreases and the CPU utilization of a web server drops below 60% and web servers are deleted, traffic will be load balanced between the reaming web servers.

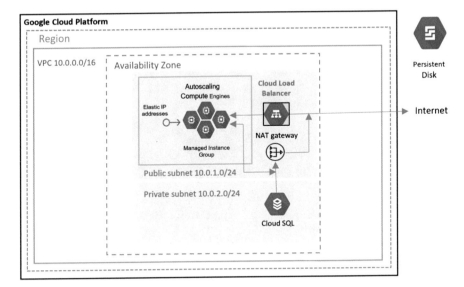

In the search area enter "load" and then select load balancing from the dropdown menu. Click on **LOAD BALANCING.**

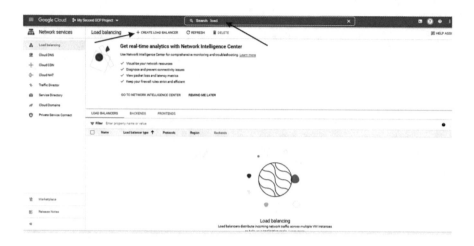

When the screen below appears, click on **START CONFIGURATION** under "HTTP(S) Load Balancing."

On the next screen enable the following:

Internet facing or internal: From Internet to my VMs or serverless service.
Global or Regional: Global HTTP(S) load balancer.

Click CONTINUE

On the next screen make the following changes:

Name: web-lb
Backend configuration highlighted: CREATE A BECKEND SERVICE

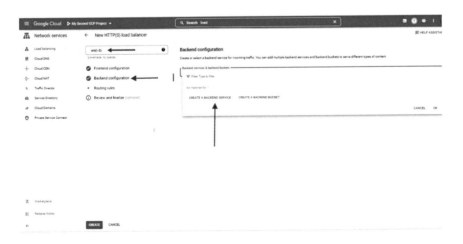

A half screen will pop-up. Enter the following information:

Name: web-backend
Backend type: Instance group
Protocol: HTTP
Named port: http
Timeout: 30
Backends:

> **New backend:**

>> **Instance group:** Instance-group-1
>> **Port numbers:** 80
>> **Balancing mode:** Utilization
>> **Maximum backend utilization:** 80
>> **Scope:** per instance
>> **Capacity:** 100

Cloud CDN: enable
Continue to scroll down the page and enter the following information.
Cache mode: Cache static content
Client time to live: 1 hour
Default time to live: 1 hour
Maximum time to live: 1 day
Cache key: Default
Health check: hc1

Logging/Enable logging: enable
Security: Blank in both
Click **CREATE**

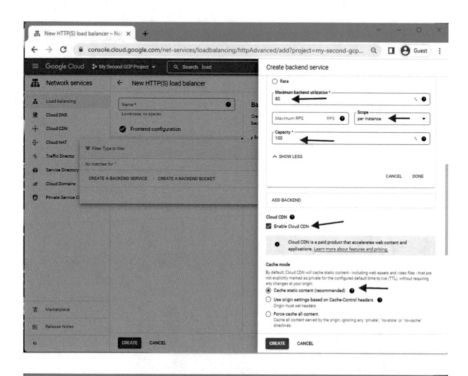

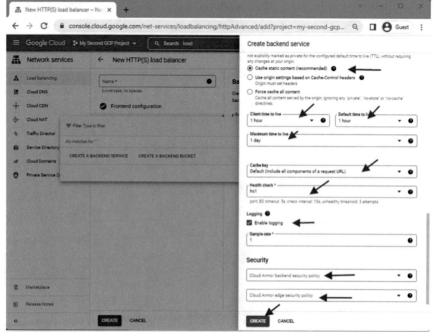

Click on **Frontend configuration.**
Name: **web-frontend**
Leave all the rest as default
Click **Done**

Click **Review and finalize.**
Click **CREATE**

Click on the **Navigation menu** icon in the upper left corner of the dashboard and scroll down to **Network services** and select **Load balancing.** The **web-lb** load balancer should appear as configured and operational.

Task 9: Create a Cloud SQL Database Instance

This task creates a Cloud SQL database based on MYSQL and connects it to **my-vpc**. Notice there is no Internet Gateway as in AWS. The Internet Gateway is integrated into each GCP VM.

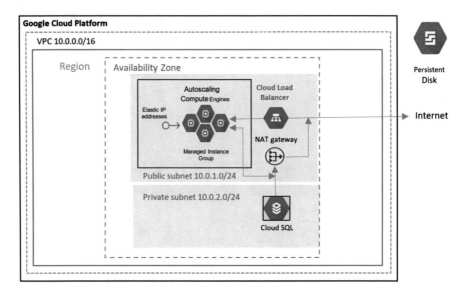

Type in **Cloud SQL** in the search bar. When the Cloud SQL Instances page appears click on **CREATE INSTANCE**.

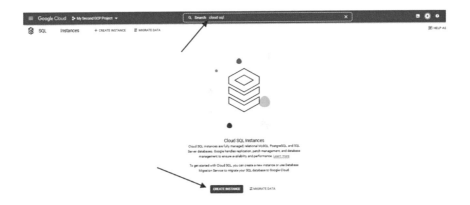

The **Create an instance** page appears. Under **Choose database engine** click on **Choose MySQL**

When the next screen appears provide the following information:

Instance ID: db-instance
No password: Enable
Database version: MySQL 5.7
Choose a configuration to start with: Production

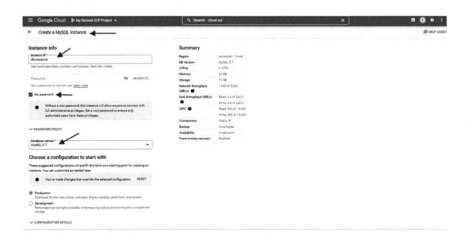

Continue to scroll down the screen.

Choose region and zonal availability:

Region: us-central1 (Iowa)
Zonal Availability: Single zone

Customize your instance:

Machine Type: High memory
Select "4 CPU, 26 GB"

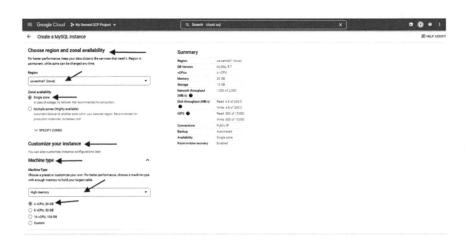

Expand **Storage:**

> **Storage type:** SSD
> **Storage capacity:** 10 GB
> **Enable automatic storage increases:** Enable

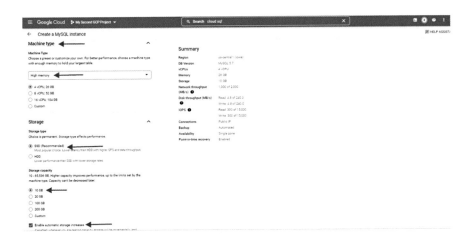

Expand **Connections:**

> **Private IP:** Enable
> **Network:** my-vpc
> **Private services access connection required:** Click **SET UP CONNECTION**
> to configure allowing the database to connect to my-vpc.

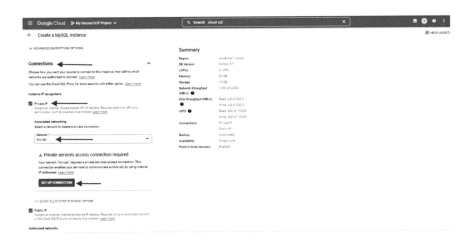

A half screen will appear.
Enable Service Network API: Enable
Allocate an IP range: click "Use an automatically allocated IP range."
Click **CONTINUE**

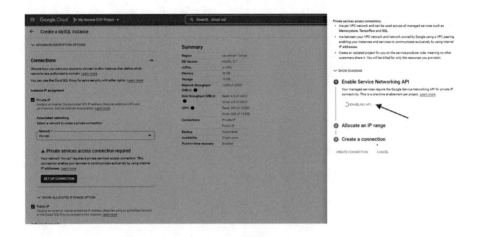

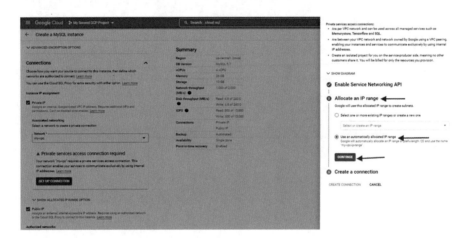

Under **Create a connection** confirm the IP range has been allocated to my-vpc.
Click **Create connection**.

When completed, the database will be connected to my-vpc.

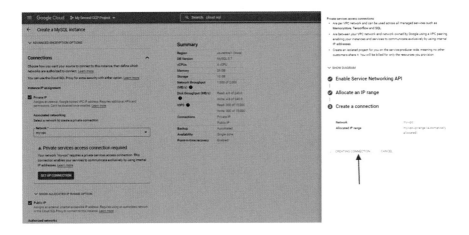

When completed you will be returned to the **Create a MySQL Instance** page. Confirm the following entries:

Private IP: Enable
Network: my-vpc
Public IP: Disable

Scroll down to the bottom of the screen and select **CREATE INSTANCE**.

The database instance will be created. The process will take several minutes. When completed the screen will verify the database has been created.

Task 10: Creating a Cloud Storage Bucket

This task will create a Cloud Storage object-based persistent disk Compute Engines can access. Notice that Cloud Storage does NOT reside in my-vpc. Cloud Storage is a global service. If allowed, it is accessible from anywhere via the Internet.

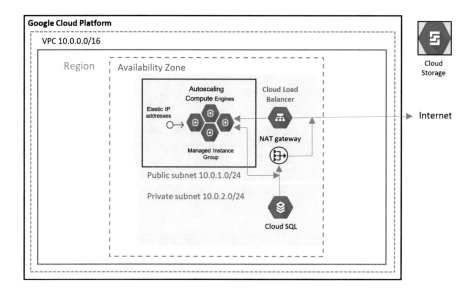

Enter "storage" in the search field and click on "Cloud Storage" in the drop-down menu.

Click **CREATE BUCKET.**

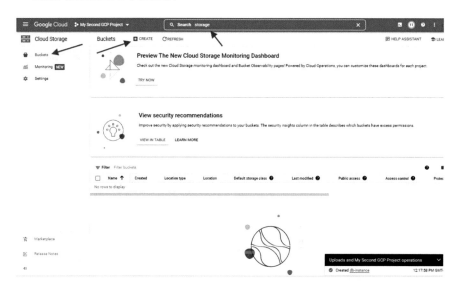

Name your bucket. The name selected must be unique across all of GPC. Note the four categories that follow:

- Choose where to store data
- Choose a default storage class for your data
- Choose how to control access to objects
- Choose how to protect object data

Click on **Choose where to store data** and provide the following information:

Name your bucket: bucket_"your name"
Location type: Region
Location: us-central1 (Iowa)
Click **CONTINUE**

Click on **Choose a storage class for your data** and provide the following information:

Enable **Set a default class and** Click **Standard**. Click **CONTINUE**

Click on **Choose how to control access to objects** and provide the following information:

Prevent public access: Enable "Enforce public access to this bucket."
Access control: Enable "Uniform."
Click **Continue.**

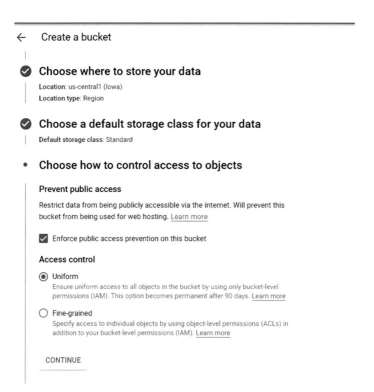

Click on **Choose how protect object data** and provide the following information:

Protection tools: None
Click **CREATE**.

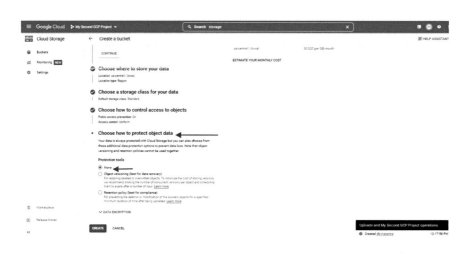

The **Bucket details** screen below will appear to verify the bucket has been created.

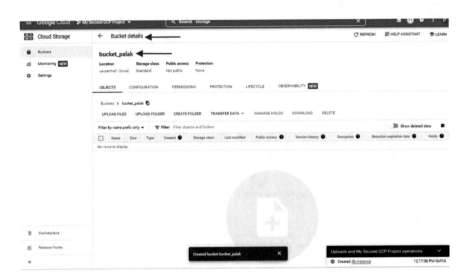

Test your bucket by uploading a file.
Click on **Upload files** and select a file to upload.
See the file that has been uploaded and download it.

11.5   Conclusion

This concludes the basic configuration of a scalable and load balanced three-tier web server architecture in the Google Cloud Platform (GCP). It is intended to be a "test drive" demonstrating basic GCP operation to allow the student hands-on experience with enough exposure to help them decide on whether to pursue further study of GCP or another public cloud provider.

In practice, this configuration is not advised. In addition, high availability should be enabled by duplicating services and data across redundant available. These and other topics should be investigated if the student decides to further study GCP.

11.6   Deletion Process

Delete Cloud Storage bucket:
Storage > Cloud Storage
Select the account(2) and click on delete (3) as shown in below picture. Once the pop-ip comes, type as suggested, here "DELETE" to confirm the deletion process.

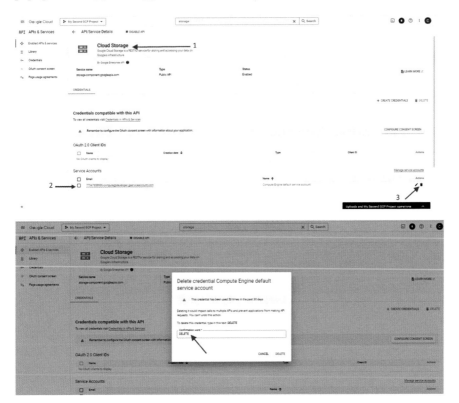

Delete Load balancer:
Network Services > Load Balancing

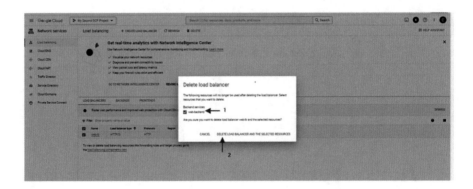

Delete Compute instance group:
Compute Engine > Instance groups
Select the **instance-group-1** and select **delete**. Confirm the delete process by typing
"**delete**" in pop-up window.

Select the **Health checks** from left pane and then select **hc1** and select delete. Confirm the delete process by typing "**delete**" in pop-up window.

Delete Cloud NAT:
Network Services > Cloud NAT
Select **nat-gateway** and click **delete**.

Delete VPC firewall rules:
VPC network > Firewall
Under **Network firewall policies**, select policy name "**allow-all**" and delete it as shown below.

Delete Virtual Private Cloud (VPC) network:

Under VPC networks, select "**my-vpc**", when the new window appears, select "private subnet" and delete it then select "public-subnet" and delete it. You will see the trash sign on right most side. And the delete "DELETE VPC NETWORK". Confirm any pop-up window by selecting delete.

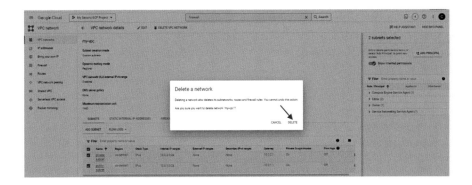

Delete Create a Project:

Go to Navigation Menu > IAM & Admin > Settings

Click on " SHUT DOWN". This will shut down the project "**My Second GCP Project**" that we created.

It will ask you to type the "Project ID", type it as it is already mentioned in the same pop-up window. Refer below picture for reference.

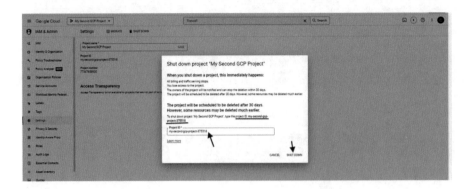

Part V
Conclusion

You have just finished hands-on labs for each of the three largest public cloud providers:

- Amazon Web Services (AWS) Lab
- Microsoft Azure Lab
- Google Cloud Platform (GCP) Lab

We have covered a lot of material. Having completed the labs for each, we now finalize comparing AWS, Azure, and GCP cloud providers. You now have the knowledge and experience to make the decision on which platform to continue studying. The path to cloud certifications are covered next followed by a summary of important cloud technology trends .

The following chapters are included:

- *Chapter 12: Comparing AWS, Azure, and GCP*
- *Chapter 13: Cloud Certifications*
- *Chapter 14: Cloud Trends*

Chapter 12
Comparing AWS, Azure, and GCP

12.1 Introduction

At this point you have learned about services offered by AWS, Microsoft Azure, and GCP and have completed hands-on labs implementing them in a three-tier website architecture. You may have a favorite, one you prefer to use. The one you choose to focus on will also determine to a large extent what type, size, and market of company you want to be employed by. Also, you should carefully consider the skillsets that each requires and decide if you are willing to learn technologies and tools used by the provider (for example, Kubernetes for GCP or Active Directory for Microsoft) to ramp up to their expectations.

When comparing AWS, Azure and GCP it appears at a high level they are all very much the same. Compute, storage, database, and networking services features and functions of each are similar. However, public cloud providers focus on specific markets. Whereas each may be able to meet any customer's needs, there are areas where one provider performs much better than the others. An analogy is a bulldozer could drive you to work, but a Cadillac is obviously the better solution for this situation.

When considering which cloud provider to pursue for certification and as a career you should compare and evaluate the following areas:

- Market share and strategy
- Use cases
- Networks and network performance
- Virtual Private Cloud (VPC's)
- Complexity
- Pricing and billing
- Customer support

© The Author(s), under exclusive license to Springer Nature Switzerland AG 2024   381
M. S. Kingsley, *Cloud Technologies and Services*, Textbooks in
Telecommunication Engineering, https://doi.org/10.1007/978-3-031-33669-0_12

- Cloud security
- Service lifecycles
- Multicloud

12.2   Market Share and Strategy

AWS entered the public cloud provider market in 2004 with IaaS. Microsoft followed in 2010 at first with PaaS but waited until 2013 to enter the IaaS market. GCP actually started in 2008 with PaaS and later with IaaS.

With a big head start, AWS has always been and is still the largest provider with 34% of the global market share. Microsoft Azure is still behind by about 21%. Even though GCP is the third largest provider, they still only have 11% of the global market. The other 40% is spread across dozens of other providers (Richter, 2022).

Both Microsoft Azure and GCP are steadily gaining market share from AWS. However, the revenue of all cloud providers is increasing with a Compound Annual Growth Rate of 17.9% resulting in a market value of $585B in 2022, reaching $1.24 T by 2027 (Cloud Computing, 2023). The percentage of the pie that each provider gets is increasing, but so is the size of the pie.

Competition is intense in the public cloud provider market. A major technical innovation could radically and quickly change the ecosystem balance. As the focus here has been on the top three current providers, there are dozens of other public cloud providers that could leapfrog to the top.

Each of the top three companies can provide services to meet most needs. However, each has strategically defined themselves to target specific customer markets. Knowing a provider's market strategy is important to be able to make the best career choice that will also complement your previous experience, interests, and goals.

12.3   Use Cases

Public cloud provider marketing strategies vary. Use cases for ideal solutions are briefly presented below for AWS, Azure, and GCP.

12.3.1   AWS

AWS focuses on medium to very large enterprises with over 200 services that can operate on multiple platforms. Backup, storage, big data, enterprise IT, websites, mobile, social, media and entertainment, and government applications are a good fit for AWS. Standout services for AWS are deep learning and Machine Learning.

A strong and diverse customer base, versatility, number of services, scalability, and reliability due to large number of Availability Zones are all major advantages of AWS.

Major AWS customers include Netflix, Twitter, ESPN, LinkedIn, Airbnb, and Facebook (Suryawanshi, 2020).

The retail industry is shy to use AWS because they see AWS, due to its connection to the retail business Amazon.com, as a competitor (Schwartz, 2018).

12.3.2 Microsoft Azure

Microsoft has leveraged their expertise with enterprise customers' as well with their existing products when designing Azure. Like AWS, Azure offers over 200 services.

The ideal customer for Azure uses Microsoft products such as SQL and Microsoft Server, Active Directory, and. NET. Azure integrates easily with Microsoft products but not well with non-Microsoft applications on servers running non-Microsoft operating systems.

A standout service for Microsoft Azure is its extensive regulatory compliance capabilities (Azure, 2023).

Major advantages of Azure include a smooth hybrid transition strategy and the largest number of data centers of any cloud provider allowing them to be located closer to customers (Fitri, 2022).

Industry segments served by Azure include retail, manufacturing, travel, financial services, healthcare, energy, and government. Azure customers include Ebay, Boeing, Samsung, GE Healthcare, Travelocity, and BMW (Leyton, 2022).

12.3.3 Google Cloud Platform

Traditionally, GCP does not have the services depth of AWS and Azure, but they are catching up. However, the GCP strategy is much different from AWS and Azure.

Small and medium-sized businesses are GCP's primary focus including an extensive startup company outreach plan. Companies that are developing cloud-native applications are also a perfect fit for GCP. They have also been successful in acquiring the business of several large telecommunications companies. Open source connectivity, which will allow integration with many other openly accessible applications, is thoroughly embraced, as is DevOps. GCP also excels in container technology, given the company's development of the widely accepted Kubernetes cloud orchestration system and the open source Machine Learning platform TensorFlow. Standout applications include big data, Machine Learning (ML), and data analytics (Gillin, 2021).

GCP customers include Home Depot, Etsy, UPS, PayPal, and Verizon (Google Cloud Customers, 2023).

12.4   Network and Network Performance

Amazon, Microsoft, and Google all have extensive global fiber optic networks. However, the Google network is considered superior because it has been designed to carry traffic from other Google applications such as their search engine. Some estimate that 25% of the global Internet traffic is carried over the Google network (Sloss, 2018).

When the customer's data reaches the public cloud provider, it is usually carried over their private fiber optic backbone network. Links generally have bandwidth of 100 Gbps (Giga-, or billion, bits per second). For instance, if a VPC needs to store data in object storage which may be some physical distance from the VPC, perhaps even in another Availability Zone, the data would transfer across the private fiber optic network, never using the Internet. There is one exception, though.

GCP has two classes of their network connectivity. With *standard* network service cloud-based connections are accomplished across the public Internet. This should actually be called "substandard" service since AWS and Azure will only carry your data inside their clouds using their private networks. Standard service is subject to delay and security problems and should not usually be used. For GCP to provide private network connections requires choosing *premium* network service at a higher cost which is standard with AWS and Azure (Network Service Tiers, 2023).

12.4.1   Availability

Availability is a measure of system uptime and most commonly used to define network availability.

Historically, telephone networks were designed to be highly available because they were everyone's lifeline to emergency services. Regulators demanded that telephone networks be designed for 99.999% (also called "five nines") of availability which is only 5 min a year of downtime (Table 12.1).

Due to the transition from traditional phone calls to those based on Voice over IP (VoIP) and different regulatory requirements that apply, the availability of many telephone providers such as cable companies is far less that that required of wireline networks. For instance, one cable provider is adverting their network is "99% (two nines) reliability" when they are actually referring to availability. As you can see from the chart, two nines availability allows almost 4 days of downtime a year (Reynolds, 2020)!

12.4.2   Durability

Data *durability* describes the average annual expected loss of stored objects. In other words, assuming I can get to the public cloud provider facilities, how sure am I that I can get my data when I want it? Typical durability for data stored in a public

Table 12.1 Annual availability vs. downtime

Number of "9's"	Percent of uptime	Downtime per year	Downtime per day
One	90%	36.5 days	2.4 h
Two	99%	3.65 days	14 min
Three	99.9%	8.76 h	86 s
Four	99.99%	56.2 min	8.6 s
Five	99.999%	5.25 min	0.86 s
Six	99.9999%	31.56 s	8.6 m

cloud provider is 99.999999999%, or eleven nines of durability. This mean that every year there is a 0.000000001 percent chance of losing a single object. For example, if you store 10,000 objects you can expect on average to lose a single object once every 10,000,000 years. Eleven nines are the expected durability provided by all major public cloud providers (Noer & Moulton, 2021). However, Azure Geo-zone-redundant storage (GZRS) provides a durability even higher of 16 nines (Morehouse, 2019).

With a durability of 11 or more nines the chances of losing data in a public cloud provider with geographic redundancy is very low. However, a durability of 100% is statistically impossible to achieve. AWS and other cloud providers have on rare occasions suffered simultaneous outages or other events, including human error, where customer data was lost.

While durability percentages for public cloud providers are extremely high, in comparison cloud provider availability percentages are very low, typically four nines for services with redundant systems and three nines for non-redundant services. Why is there such a difference?

To reach a public cloud provider requires connecting to a public network like the Internet or a private network for a provider such as AT&T or Verizon. The public Internet provides no guarantees of service quality. It is prone to delays and outages. On the other hand, if you use the services of a private network you can usually count on much improved network availability, even five nines for dedicated network connections. However, the public cloud provider is not going to accept much responsibility for your data getting to their cloud since they do not have control over the connections. Therefore, the discrepancy between the network availability and the data durability.

12.4.3   Service Level Agreements and Contracts

Public cloud providers define their commitments to services performance in Service Level Agreements (SLA's). SLA's are contracts between the cloud provider and the customer regarding availability and other metrics such as latency. When doing a search on a cloud provider's SLA's you will find there a dozens, if not hundreds, of SLA documents specific to different services.

Contracts, including SLA's, are written to favor the public cloud provider. Due to customer pressure, this trend is slowly changing. Even though SLA's are legally binding, the burden of proof is on the customer to prove they have not been met which requires expensive monitoring systems and skilled people to operate then and interpret the results. Further, should proof be presented, the compensation for downtime is minimal and the cloud provider is not responsible for related damages like loss of revenue due to network delay or unavailability.

Contracts for service with a cloud provider can be negotiated depending on how much money you will spend with the provider. However, don't expect to bargain with the provider if you are only spending a few thousand dollars. The more you will spend with them the better terms that can be negotiation.

Public cloud provider contract negotiations usually require the assistance of lawyers. There are limits on what can be negotiated. For instance, public cloud providers know what SLA's they can offer and still consistently meet them. They are not going to offer SLA's that put them at risk of not delivering as expected. They may, though, help you design a solution that can increases redundancy but that will certainly be more expensive.

12.5   Cloud Provider Virtual Networks

A virtual network is actually networking in data center servers. Their capabilities should be evaluated when considering a public cloud provider.

12.5.1   AWS VPC

AWS calls their virtual network entity a Virtual Private Cloud (VPC). Each VPC belongs to an AWS region. VPC's operate on public and private subnets that are confined to Availability Zones which are distinct locations isolated from failures in other availability zones. All resources inside the VPC can communicate with each other.

12.5.2   Azure VNET

Azure's virtual network entity is called a VNET. VNET's are confined to one Azure region. Subnets are assigned to regions. All resources on subnets can communicate with all other resources including those in different subnets by default. Unlike AWS, instances in Azure have Internet access by default, no Internet Gateway is required. Also, route tables are configured automatically rather than manually by the user in AWS.

12.5.3 GCP VPC

Like AWS, GCP's virtual network entity is also called a Virtual Private Cloud (VPC). Even though GCP is based on regions, the GCP VPC is global spanning all regions and not belonging to any specific individual region. When a VPC is created, subnets are also created in all regions by default. Those subnets are region specific. By default all instances can communicate with all others in the network automatically.

12.6 Complexity

Cloud provider infrastructures are complex but different public cloud providers are better or worse depending on the provider. Knowing these differences should be considered before choosing a public cloud provider.

12.6.1 Services

A large number of services results in a lot of versatility for customers but also to complexity. In a rush to add more services, integrating those services together is often neglected. However, Azure is the leader in integrating services together for seamless operation (Prewett, 2020).

12.6.2 Documentation

Public cloud providers are large organizations that have thousands of customers. Getting timely support can be difficult and often requires paying a lot for it. Therefore, good documentation is essential so the customer can solve some of their problems on their own. AWS is credited with excellent documentation systems while Azure and GCP have been harshly criticized for poor documentation of services (Clinton, 2019; Compton, 2018; Microsft, 2022).

12.7 Pricing and Billing

Pricing of services is especially complex, particularly for AWS. With over 200 services, pricing is often determined on an individual basis with many levels depending on how the service is used. In addition, many different saving plans can be applied resulting in even more confusion. This array of options flows downstream to the billing system.

To AWS's credit, they do offer many tools to help the customer control their usage and to understand their billing (Amazon, 2023). But once again, with dozens of tools available it requires a high level of expertise to apply them.

Azure and GCP provide more streamlined pricing structures than AWS. However, Azure is more expensive and data downloads from GCP are very expensive in comparison to AWS.

12.8   Customer Support

As a customer of a large cloud provider you will have very limited customer support by default. However, you can contract with all three providers for extended support. Commitment levels range from minimal telephone support to being assigned to an engineering team with guaranteed response times in minutes. Of course, this level of support is expensive and may not be affordable.

Azure customer support is considered at best subpar and at worst terrible (MetrixData360, 2022).

12.9   Cloud Security

Security capabilities vary widely across cloud providers. All providers adhere to the Shared Responsibility Model. In practice, security applications used in a conventional data center may not be usable in a cloud-based infrastructure.

12.9.1   AWS

One of the biggest advantages of AWS is their depth of security experience. However, their security tools evolved quickly and often appear to have been "kludged" together.

One way to describe AWS security is "locked down." Services generally default to secure configurations but a service cannot be accessed unless it is explicitly enabled. IAM is a mature and effective product. However, having to explicitly enable services with manual IAM operation can dramatically hinder timely scaling of capabilities.

AWS does have the best core security services available. Two good examples are well designed security group features and very granular IAM (Prokopets, n.d.).

12.9.2   Azure

Of the three providers, Azure has the worst cloud security strategy. This is primarily due to the inconsistency in engineering security features. Often two features do not respect the limitations of each other. These errors cannot be fixed by the customer and are often ignored by Microsoft. (Quinn, 2022).

In contrast to AWS, Azure cloud security is not "locked down," in many cases security is left completely open in Azure. For example, a VPC in AWS (as well as GCP) is created in the "default deny" mode on ports and protocols while in Azure they start as "default open."

Azure does do some things right in securing their cloud. Active Directory is the centralized Identity and Access Management system for Azure that allows centrally configuring users, groups, and so on whereas AWS IAM must do so in separate steps.

Another advantage is Azure activity logs are collected by default across regions. AWS requires a solution such as creating a Lambda function to accomplish this.

Finally, Azure allows subscription level access so individual teams can respond to alerts that affect them.

Unfortunately, after Active Directory, activity logs, and subscription level access for alerts, cloud security in Azure cloud security is unwieldly.

12.9.3   GCP

In the middle of the pack in regards to cloud security is GCP.

Similar to AWS, GCP has granular IAM and defaults to secure configurations and like Azure has organization-wide logging. However, GCP lacks the depth of security features AWS offers but does have some tools that have impressive features but some are still in beta testing mode.

A huge advantage for GCP is they are engineering experts, as demonstrated by their dominance in container management and AI, and are definitely experienced on a global level as the result of their other products like Google Search. They are quite capable of expanding their cloud security capabilities.

A downside is there is a very limited number of qualified GCP Security experts available to be employed by GCP customers.

12.10   Services Lifecycle Strategy

Services have a lifecycle that starts with the introduction of the service and ends when the service is discontinued. Once customers are using a service they are hesitant to stop using it. Customers also want full backward compatibility of new service versions with older ones.

Both AWS and Azure have acceptable lifecycle policies and backward compatibility standards. However, GCP is known for having a less than acceptable "deprecation," or end of service life, standards which may adversely affect your GCP operation more than if you were an AWS of Azure customer.

12.10.1 Caution

A disturbing event occurred in 2020 where AWS abruptly removed the website Parler from their platform. The reasons were obviously because AWS disagreed with the political motivation behind the website. Ignoring the free speech, moral, or Constitutional issues that will undoubtedly be debated for a long time, this event brings up some very basic business concerns.

Some would naively assert that Parler could have simply moved their website to another cloud provider. Actually, this is not as easy as it may appear. Once a business integrates their data and applications onto the cloud providers hardware and services, it cannot just be pulled out and ported directly to another cloud provider. There is a lot of complicated, time-consuming, and expensive work to move to another cloud provider and their services. This may not be feasible for some businesses.

Let's assume that if a business does not generate revenue for 90 days they will be out of business. For most it is less time than that. When AWS removes a customer from their platform they don't just inconvenience the business, in all likelihood they end it.

This situation is an unanticipated risk factor that did not exist before the Parler incident. When considering migrating to a public cloud provider the culture and political undercurrents of your company now must be considered before making the decision to move to a public cloud.

12.11 Multicloud

Now that we have compared the "big three" cloud providers as though one would be declared the winner, let's consider a more likely scenario. Most large enterprises do not use the services of just one cloud provider. It is not uncommon for some very large organizations to connect to as many as a dozen or more different cloud providers. This arrangement is now being called the **multicloud**.

Let's look at an example. Suppose your company has a large e-commerce website, runs a lot of Microsoft applications, and uses Kubernetes. Of course you might be able to meet all these needs with one cloud provider, but the best solution in this case would be to use AWS, Azure, and GCP due to their individual strengths.

Another reason to use a multicloud is redundancy. Either splitting data between two or more cloud providers or running duplicate applications in two both to reduce risk of losing data or not being able to access applications if a cloud provider fails are both good alternatives.

Customers are concerned about vendor lock-in where a cloud provider may in the future change their services or business strategy that is counter to your organization's goals. Having at least one other provider can hedge that risk.

Finally, using more than one cloud provider can provide the company with financial leverage when negotiating cloud services between competing providers.

12.12   Summary

When considering which public cloud provider to use for service or to pursue as a career it is important to analyze where the provider excels, their market strategy, their customers, and their corporate culture. You will find that each public cloud provider has a niche where they fit best.

AWS, Azure, and GCP each has advantages and disadvantages and choosing one over the others can be a difficult decision. That decision should be made based on customer requirements, your priorities, and perceived risks and by comparing those to the capabilities of the provider.

In general, though, AWS has historically offered more services and has the most experience. If your organization has extensive Microsoft products then they would be favored. Finally, GCP has positioned itself in the market as the lowest cost provider, for providing support for startups, and for embracing open source software.

Below is a summary and comparison of the major AWS, Azure, and GCP services (Table 12.2).

Table 12.2 Comparison of AWS, Azure, and GCP services

	AWS	Azure	GCP
Compute	Elastic Cloud Compute (EC2)	Azure Virtual Machines	Compute Engine
Identity and access	Identity and Access Management (IAM)	Azure Active Directory	Cloud IAM
Virtual network	Virtual Private Cloud (VPC)	Vnet	Virtual Private Cloud (VPC)
Load balancing	Elastic Load Balancing (ELB)	Azure Load Balancing	Cloud Load Balancing
Automatic Scaling	Auto Scaling	Auto scale with VM Scale Sets	Instance Groups
File storage	Elastic File System (EFS)	Azure Files	Filestore
Block storage	Elastic Block Storage (EBS)	Azure Disk Storage	Persistent Disk
Object storage	Simple Storage System (S3)	Azure BLOB Storage	Cloud Storage
Archive storage	Glacier	Azure Cool Storage	Nearline and Coldline Storage
Relational database	Relational Database Service (RDS), Aurora	Azure SQL Server	Cloud SQL
Nonrelational database	DynamoDB	CosmoDB	BigTable/Datastore
Serverless	Lambda	Azure Functions	Cloud Functions

Homework Problems and Questions

1.1 Which criteria should be considered when considering a public cloud provider?
1.2 Compare and contrast the strengths and weaknesses of AWS, Azure, and GCP.
1.3 Which public cloud provider would be the best solution for your organization?

Bibliography

Amazon. (2023). *Cloud financial management.* [Online] Available at: https://docs.aws.amazon. com/whitepapers/latest/aws-overview/aws-cost-management.html
Azure. (2023). *Azure compliance.* [Online] Available at: https://azure.microsoft.com/en-us/ explore/trusted-cloud/compliance/
Clinton, D. (2019). *Why is so much enterprise documentation so awful?.* [Online] Available at: https://www.freecodecamp.org/news/why-enterprise-documentation-awful/
Cloud Computing. (2023). *Cloud computing market.* [Online] Available at: https://www.market-sandmarkets.com/Market-Reports/cloud-computing-market-234.html
Compton, D. (2018). *Google cloud platform – The good, bad, and ugly (It's mostly good).* [Online] Available at: https://www.deps.co/blog/google-cloud-platform-good-bad-ugly/
Fitri, A. (2022). *Where are the hyperscale cloud providers building their data centres?.* [Online] Available at: https://techmonitor.ai/technology/cloud/where-cloud-providers-building-data-centres
Gillin, P. (2021). *Special report: Google cloud rising.* [Online] Available at: https://siliconangle. com/2021/10/11/fits-starts-googles-cloud-strategy-finally-finding-footing/
Google Cloud Customers. (2023). *Google Cloud.* [Online] Available at: https://cloud.google.com/ customers
Leyton, D. (2022). *Which companies are using Microsoft Azure?* [Online] Available at: https:// www.websitebuilderinsider.com/which-companies-are-using-microsoft-azure/
MetrixData360. (2022). *Microsoft Azure: The Pros and Cons.* [Online] Available at: https://metrix-data360.com/cloud-series/microsoft-azure-the-pros-and-cons/
Microsft. (2022). *Why does azure documentation suck so bad?* [Online] Available at: https://www. teamblind.com/post/Why-does-azure-documentation-suck-so-bad-YvvM7iPp
Morehouse, J. (2019). *Understanding Azure geo-redundant storage.* [Online] Available at: https:// sqlrus.com/2019/12/understanding-azure-geo-redundant-storage/
Network Service Tiers. (2023). *Empowering customers to optimize their cloud network for perfor-mance or price.* [Online] Available at: https://cloud.google.com/network-tiers
Noer, G., & Moulton, D. P. (2021). *How cloud storage delivers 11 nines of durability – And how you can help.* [Online] Available at: https://cloud.google.com/blog/products/storage-data-transfer/ understanding-cloud-storage-11-9s-durability-target
Prewett, M. (2020). *Microsoft Azure Benefits (in-depth): Benefit #1: Top-notch Synergy Between Services.* [Online] Available at: https://key2consulting. com/5-microsoft-azure-benefits-and-challenges/
Prokopets, M. (n.d.). *AWS Security vs. Azure security: Cloud security compariso.* [Online] Available at: https://nira.com/aws-security-vs-azure-security/
Quinn, C. (2022). *Azure's security vulnerabilities are out of control.* [Online] Available at: https:// www.lastweekinaws.com/blog/azures_vulnerabilities_are_quack/
Reynolds, R. (2020). *Achieving "five nines" in the cloud for justice and public safety.* [Online] Available at: https://aws.amazon.com/blogs/publicsector/achieving-five-nines-cloud-justice-public-safety/

Richter, F. (2022). *Amazon, Microsoft & Google Dominate Cloud Market.* [Online] Available at: https://www.statista.com/chart/18819/worldwide-market-share-of-leading-cloud-infrastructure-service-providers/

Schwartz, S. (2018). *Dramatic or justified? Retailers' fears push cloud customers from AWS to Microsoft, Google.* [Online] Available at: https://www.ciodive.com/news/dramatic-or-justified-retailers-fears-push-cloud-customers-from-aws-to-mi/543273/

Sloss, B. T. (2018). *Expanding our global infrastructure with new regions and subsea cables.* [Online] Available at: https://cloud.google.com/blog/topics/inside-google-cloud/expanding-our-global-infrastructure-new-regions-and-subsea-cables

Suryawanshi, N. (2020). *The biggest AWS users.* [Online] Available at: https://www.linkedin.com/pulse/biggest-aws-users-nikhil-suryawanshi/

Chapter 13
Cloud Certifications

13.1 Congratulations!

If you are here, you have studied the textbook material and completed the labs (unless you looked ahead!). You now have sufficient understanding of cloud and related technologies as they relate to Amazon Web Services (AWS), Microsoft Azure, and Google Cloud Platform (GCP) public cloud providers' platforms. Hopefully, you have selected the one you want to study further. I am sure you are wondering what the next step is.

13.2 Cloud Certifications

Even if you have vast industry experience or advanced degrees, the pathway to validation of your knowledge is through public cloud provider certifications. The focus of this textbook has been on "solutions architect" or "cloud architect" certifications. However, as mentioned previously, although network and cloud areas are still considered separate certifications by cloud providers, they are beginning to be considered together by employers, so you will also see reference to network/cloud architect in job descriptions. Cloud architects are being required to cross train in networking, and network engineers are expected to expand their qualifications in cloud technologies.

You may have already looked at the "big three" certification pathway websites and discovered there are many other certifications available, as well as specialty or advanced certifications. You may consider advancing your knowledge and skills in topics such as artificial intelligence, machine learning, python programming, advanced networking, and many more. For now, let's concentrate on the Cloud Architect pathway. Details of AWS, Microsoft Azure, and GCP follow.

M. S. Kingsley, *Cloud Technologies and Services*, Textbooks in Telecommunication Engineering, https://doi.org/10.1007/978-3-031-33669-0_13

Fig. 13.1 AWS Solutions
Architect Path

13.2.1   AWS Solutions Architect Path

AWS has three levels of architect certifications: Cloud Practitioner, Solutions Architect Associate, and Solutions Architect Professional (Fig. 13.1).

13.2.1.1   AWS Cloud Practitioner

The AWS Solutions Architect path begins with the Cloud Practitioner certification exam. It validates that you have foundational skills in the AWS cloud. AWS recommends the following before attempting the Practitioner exam:

- Six months of experience with the AWS cloud. This can be acquired through various AWS and independent study and training opportunities.
- AWS use cases and how AWS can improve a business.
- Basic understanding of IT and how it is applied in AWS.
- Knowledge of AWS core services (compute, storage, databases, and networking) and how they are implemented.
- Billing and service pricing for AWS services.
- Basic cloud security concepts.

The Cloud Practitioner examination has 65 multiple-choice or multiple-response question with a time limit of 90 minutes. A passing score is typically 72%.

13.2.1.2   AWS Solutions Architect Associate (SAA)

The AWS Solutions Architect Associate certification is one of the most coveted certifications. It will validate your ability to design and implement distributed systems on AWS. This exam is comprehensive and difficult. It will require extensive study as well as the following recommended experience:

- One year of hands-on experience implementing compute, storage, databases, networking, and related support systems. As with the practitioner certification, this experience can be achieved via various training and educational channels or through employment.
- Ability to implement security and compliance controls for workloads.
- Understand and apply the AWS Well-Architected Framework (WAF).
- Understand the AWS global infrastructure.

- Evaluate network requirements and deploy their solutions.
- Implement security safeguards at multiple levels of the AWS architecture.

The AWS Practitioner certification is not a prerequisite for the Solutions Architect examination.

The Solutions Architect Associate examination has 65 questions and a time limit of 130 minutes. Unlike the practitioner exam, all questions are scenario-based. You will be given complex problems and are expected to derive the *best* solution, often from choices that are all correct.

13.2.1.3 AWS Solutions Architect Professional

The AWS Solutions Architect Professional certification requires extensive industry experience. Obviously, it will require a lot of work. However, the compensation and promotional opportunities usually do not motivate achieving it. Professionals should expect only small increases in salary over SAAs.

13.2.2 Microsoft Azure

The Microsoft Azure Solutions path has two certification steps: Azure Administrator Associate and Azure Solutions Architect Expert (Fig. 13.2).

13.2.2.1 Microsoft Azure Administrator Associate

Microsoft Azure Administrator Associate exam is usually the entry point for the Microsoft architect path with previous experience with Azure. However, the Microsoft Fundamentals exam may be taken first.

The Administrator Associate is responsible for implementing, managing, and monitoring compute, storage, virtual networking, and other services supported in

Fig. 13.2 Microsoft Azure
Solutions Architect Path

Azure. Knowledge and experience recommended before taking the Microsoft Azure Administrator Associate exam include:

- Six months of hands-on experience administering Azure. Experience can be gained via many avenues, including employment, Microsoft third-party training programs, or self-study.
- Understanding of Azure core services.
- Azure workload provisioning and management.
- Azure security and governance concepts.
- Proficiency with the Azure portal
- Some experience with PowerShell.

The Azure Administrator Associate examination has 40–60 questions for which you will have 150 minutes to complete the exam. Questions will include multiple choice, drag-and-drop, case studies, or multiple response. A passing score of 70% is usually required.

13.2.2.2   Microsoft Azure Solutions Architect Expert

The Microsoft Azure Administrator Associate certification is a prerequisite for taking the Microsoft Azure Solutions architect Expert examination. In addition, passing the Designing Microsoft Infrastructure Solutions examination is required. When these requirements are met, the Azure Solutions Architect Expert certification is awarded.

Experience and knowledge recommended before taking the Designing Microsoft Infrastructure Solutions examination include the following:

- At least 1 year of hands-on experience using Microsoft Azure.
- Advanced knowledge of Azure compute, storage, databases, and networking components for implementing identity, business continuity, governance, and security solutions.
- Ability to design solutions to meet business objectives.
- Expertise in Azure administration, development, and DevOps processes.

The Designing Microsoft Infrastructure Solutions examination has 40–60 questions for which you will have 150 minutes to complete the exam. Questions will include multiple choice, drag-and-drop, case studies, or multiple responses. A passing score of 70% is usually required.

13.2.3   Google Cloud Platform

GCP has two levels of architect path certifications: Cloud Engineer and Cloud Architect (Fig. 13.3).

Fig. 13.3 GCP
Architect Path

13.2.3.1   Google Cloud Engineer

An associate level certification, the Google Cloud Engineer deploys, monitors, and manages enterprise-level cloud solutions.

Before attempting the Google Cloud Engineer examination, you should have experience:

- At least one-year hands-on use of the Google Cloud Platform.
- Setting up a cloud environment.
- Planning and deploying cloud solutions.
- Monitoring and maintaining cloud operations.
- Integrating access and security into cloud solutions.

The Cloud Engineer examination has fifty multiple choice (one correct answer) and multiple select (more than one correct answer) questions with a two-hour time limit. A passing score is 70%.

13.2.3.2   Google Cloud Architect

A professional level certification, the Google Cloud Architect designs, implements, and manages secure, scalable, reliable, and redundant cloud solutions that meet business needs. Before taking the Google Cloud Architect examination, you should have the following experience:

- Designing and deploying cloud architectures.
- Create cloud solutions that meet business objectives.
- Integrate regulatory compliance into cloud solutions.
- Designing redundancy and reliability into cloud architectures.
- Create highly secure cloud environments.

The Cloud Architect examination has fifty multiple choice (one correct answer) and multiple select (more than one correct answer) questions with a two-hour time limit. A passing score is 70%.

13.3   Where to Get Training

Several options are available, depending on your aptitude and budget.

13.3.1   Public Cloud Provider Training

Public cloud provider training is outstanding. You will get training from those that create the certification exams. Combinations of live and recorded, onsite, and remote options are usually available. However, their training options are usually expensive.

13.3.2   Professional Training Organizations

Third-party organizations are plentiful and offer excellent training. Most monitor the certification exams and integrate material into their training curriculums. Just like public cloud provider offerings, third-party training is usually expensive.

13.3.3   Udemy and Coursera

Courses available from online training organizations are on par with any other training option and are an incredible value. Full length programs with extensive recorded lectures with detailed lab instruction are available from multiple sources for as little as ten dollars on Udemy and Coursera.

13.3.4   Self-Study

Training can be accomplished the old-fashioned way-but a book and study. Multiple good published sources are available. Be sure your choice has access to and detailed instructions on using public cloud providers' labs. Certification exams are revised often. Make sure your publication covers the most current revision of your target examination.

13.4   Compensation

So, the question on your mind is probably "What do I get for all this effort?" The good news is that cloud engineering is financially lucrative.

Compensation for Microsoft Azure and Google Cloud Platform are generally higher than Amazon Web Services. Google Cloud Platform compensation is highest presumably since they have smaller market share than AWS or Microsoft and must pay more to attract employees.

In general, the compensation for Cloud Engineers is within average ranges regardless of provider:

- Junior Network/Cloud Architect $100 k-$130 k
- Mid-level Network/Cloud Architect $140 k-$160 k
- Senior Network/Cloud Architect $170 k-$200 k+

As you can see, compensation for cloud architects is very good! However, do not pursue this as a career unless, during your evaluation period, you find you enjoy this work!

13.5 Summary

This chapter reviewed the cloud certifications available from AWS, Microsoft Azure, and GCP. Training opportunities are recommended. Finally, compensation expectations based on the current market for cloud engineers by cloud provider and certification are presented.

Homework Problems and Questions

1.1. Describe the certification programs offered by Amazon Web Services, Microsoft Azure, and the Google Cloud Platform.
1.2. Do you plan to pursue cloud certifications? Which one(s)? Why?

Chapter 14
Cloud Industry Trends

14.1 Cloud Trends

The cloud industry continues to evolve rapidly. Therefore, as a cloud engineer, you will be continually learning new technologies or, sooner or later, face obsolescence. Constantly studying will be your challenge!

Several cloud technologies are currently experiencing rapid changes. These changes should be closely monitored and skills acquired to meet their demands. They are the evolution of:

- Cloud services management
- Applications development
- Network architectures
- Network security
- Automation and programmability

AWS services are used as examples. However, equivalent services from Microsoft Azure and GCP are listed for comparison.

14.2 Cloud Services Management

Cloud services have evolved through three phases:

1. Unmanaged services
2. Managed services
3. "Serverless."

© The Author(s), under exclusive license to Springer Nature Switzerland AG 2024   403
M. S. Kingsley, *Cloud Technologies and Services*, Textbooks in
Telecommunication Engineering, https://doi.org/10.1007/978-3-031-33669-0_14

14.2.1 Unmanaged Services

AWS EC2 compute was an evolutionary, even revolutionary, concept. By using EC2 virtual machines, the number of physical servers could be dramatically reduced, resulting in huge cost savings for cloud operators, which, when passed on to customers along with a pay-as-you-go pricing model, made moving on-premises data center infrastructures to the cloud financially feasible.

However, EC2 is not a "managed" service. In other words, the customer must define many inputs to EC2. Parameters including CPU capability, size, processor, storage, memory, security, and operating system must be predefined as well as any redundancy requirements.

That is a lot to consider. Therefore, when using EC2, the operating environment must be well understood. Configuring too many resources will result in paying too much for service. Too little will result in poor performance. "Tweaking" of parameters may be necessary to find the right number, size, and configuration of EC2s to serve an application.

Many other services are unmanaged. However, the next advancement was services "managed" by the cloud provider relieving the customer of those duties so they can focus on developing and delivering applications.

14.2.2 Managed Services

AWS managed services shift much of the provisioning, operation, and maintenance of a service responsibility from the customer to the cloud provider. An example is the AWS RDS database using Oracle.

The long blue vertical line on the left side of Fig. 14.1 represents an on-premises, standalone database such as Oracle. The company is responsible for all areas of database operations from installation to maintenance. Alternatively, AWS can implement an Oracle database in two ways. The first is directly onto an EC2. This method allows the customer maximum access to the EC2 and the database. The short red vertical arrow on the right shows that AWS is responsible for the installation of equipment and server operating systems, supplying all power and air conditioning, and installing the operating system. The rest of the database management, represented by the vertical blue arrow on the right side, is the responsibility of the customer.

The second method is to integrate Oracle with the AWS Relational Database Service (RDS). RDS is a wrapper that connects Oracle to AWS internals. When using RDS and running Oracle AWS accepts all responsibility for all operation, management, and maintenance of RDS and Oracle. RDS is a "fully managed" AWS service. This is represented by the long red vertical red line on the right side of the figure.

Fig. 14.1 AWS databases
on-premises, managed, and
unmanaged services

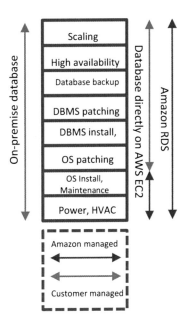

14.2.3   Serverless

The term "serverless" is misleading. Obviously, services and applications can't run on air, they require hardware servers to operate. The term serverless means the customer does not have to configure, provision, or in any way interface with hardware to implement a service. It also implied the service will be operational on-demand and only on for the amount of time it is needed. Serverless is the next step beyond managed service and is completely hands-off. It is also the next generation of the "pay-as-you-go" concept whereas the service literally does not exist until there is a request for it and it disappears after its demand is accomplished.

Serverless operates using the Function-as-a-Service (FaaS) concept (Redhat, 2020). It is an execution model where the cloud provider executes code (a function) to dynamically allocate cloud resources. The code that powers a serverless operation is actually the instructions to enable the infrastructure needed to support an application. For instance, all the instructions are needed to create an EC2. Once the code is written to turn on an EC2, the provisioning of resources is automatic, requiring no human intervention. They are allocated for only as long as they are needed to perform needed operations. When the operation is completed, the resources are not just turned off but deleted. The customer is only charged for the amount of time they are actually using the cloud provider resource.

14.2.3.1   AWS Lambda

The AWS FaaS is *Lambda*. It is completely serverless. Instead of an administrator configuring and managing, for example, a web server using EC2 compute virtual machines with load balancing and autoscaling, Lambda defines instructions to create those services by executing code.

Lambda is event-driven. In other words, the instructions to create EC2s and related services are written in code that is "triggered," or run, when initiated by an *event*, in this case a request for a web page. The concept of using code to initiate the creation of hardware is called "Infrastructure-as-Code."

Lambda can handle millions of short duration requests per second. A good example is Amazon Alexa. When a user interacts with Alexa and requests, for example, "Alexa, play a romantic song," the request triggers Lambda computer code that sets up the resources (EC2, S3, etc.) and delivers the request. It then deletes the resources.

Many Lambda functions are created, or coded, and are waiting for an event trigger to run the code and provide the designated functions.

A more detailed example is a request to update a database in a three-tier web application (Fig. 14.2).

1. A user requests a web page hosted on S3 and wants to modify their account information.
2. The user login and is authenticated using AWS Cognito.
3. The modified request is also forwarded to the API Gateway, which recognized the request as needing to access the DynamoDB database to update the user's account information.
4. The API Gateway communicates with Cognito to verify the user is authorized for this action.
5. The API Gateway sends a trigger to Lambda.
6. Lambda runs the code to access DynamoDB and update the user's account information.
7. Once the user's information is updated, connectivity to Lambda tears down all the related services including S3, Cognito, API Gateway, Lambda, and DynamoDB.

Fig. 14.2 Three-tier web application database update using Lambda

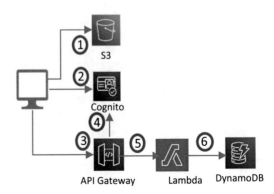

As illustrated, instead of a resource (in this case, DynamoDB) constantly running and waiting 24/7 to be used, Lambda initiates its use only when it is needed. Therefore, charges for use of resource only occurs when it is actually used. Billing is applied in millisecond increments.

Lambda is robust enough to service millions of requests per second.

Microsoft Azure: Azure Functions
Google Cloud Platform: Cloud Functions

14.2.3.2   Lambda Serverless Versus Microservices

Lambda is intended to be used for high-volume requests that can be completed quickly. The maximum time Lambda can run is 15 seconds. On the other hand, if you are building a large platform with constant scaling involved then microservices, which are introduced and explained shortly, are the best solution. However, before looking at microservices, we need to look at what other application development methods, although becoming obsolete, are still being used today.

14.3   Applications Development

Applications development has evolved to include more efficient ways of building applications. Methods include:

- Monolithic applications
- Service Oriented Architecture (SOA)
- Microservices
- Containers
- DevOps

Each of these are discussed in detail below.

14.3.1   Monolithic Applications

Applications have traditionally been developed where all of the functionality is contained in a single code base and released as a single file. Applications developed in this way are referred to as "monolithic." As the name implies, the end result is the application software size is usually large and inflexible and results in several disadvantages:

- Monolithic code can be hard to manage.
- Tight coupling between modules makes it difficult to break apart the application if needed.

- Even small changes require redeploying the entire application.
- A single bug can bring down the whole application.
- Scaling can be difficult.
- It is hard to integrate new technologies because they will affect the whole application.

However, developing monolithic applications does offer the advantage in that they are easy to develop and deploy compared to more efficient methods.

14.3.2 Service-Oriented Architectures (SOA)

Service Oriented Architectures (SOA) were developed to solve monolithic development problems. In SOA, a large application is decomposed into many smaller services that are deployed independently. These services communicated indirectly via middleware that uses various messaging standards between components (Gill, 2022).

Because services are independent units' advantages of SOA include:

- The code is easier to maintain.
- Components can be reused in other applications.
- Debugging and testing is easier so the application is more reliable.
- Development can be accomplished in a parallel effort by independent development teams.

However, SOA does have disadvantages:

- The main disadvantage of SOA is its complexity. For instance, coordinating messaging between components is difficult.
- The investment in people and technology is initially high.
- The interaction between components and related messaging can cause increased application response time and decreased performance.

14.3.3 Microservices

Microservices are an architectural approach to application development where software is composed of small independent services that communicate over well-defined APIs that together result in a comprehensive service offering (Fig. 14.3).

Microservices are an extension to SOA. In effect, microservices are fine-grained SOA, and are loosely coupled. Microservices-based applications are easier to scale and faster to develop.

A feature of the microservices architecture over monolithic is that each microservices is supported by its own database. Therefore, response times to microservices requests is fast and performance is improved.

Fig. 14.3 Microservices

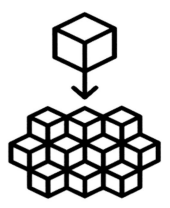

Advantages of microservices architectures include:

- **Agility.** Microservices lend themselves to development and management by small teams which act autonomously and can react quickly to changes.
- **Flexible scaling.** Individual services can be scaled independently allowing efficient and effective application sizing.
- **Easy deployment.** Microservices are adapted to continuous integration and continuous delivery (CI/CD). Changes or improvements can be easily applied and tested.
- **Decentralized.** Services can be spread across multiple systems and even across different geographies.
- **Technological freedom.** Development teams are free to use any tools that meet specific needs rather than being bound to those defined by more formal development environments.
- **Reusable code.** Individual services can be reused as part of other applications.
- **Resiliency.** In monolithic architectures, if an individual component fails the entire application will fail. Since microservices are individualized a problem with an individual service will result only in degraded operation and not complete failure of the system.
- **Improved fault isolation.** When a failure occurs in a microservices architecture, problems can be quickly isolated to individual services which can be more rapidly repaired.
- **Improved testability.** Individual services are tested independently streamlining and simplifying the testing process.
- **Improved maintainability.** Individual services are easier to maintain compared to monolithic structures.

While offering extensive advantages, microservices do present some disadvantages,

- **Complexity**. Being distributed systems, microservices are inherently more complex that monolithic applications.
- **Specialized skillsets.** Developers must adapt to microservices methods and acquire specialized skills.

- **Network cost.** Microservices can result in more costly network services to connect distributed services.
- **Security.** Also due to their distributed nature, microservices are more vulnerable to security breaches.
- **Database integration.** Each service in a microservices architecture has its own database. If migrating from a monolithic to microservices architecture it can be difficult to decompose the monolithic database for use by individual services.

Deploying microservices is best accomplished using "containers."

14.3.4   Containers

Network services are becoming commoditized. In other words, the profit from them is constantly decreasing due to extreme competition between providers. Network and cloud service providers generate the most revenue from applications. For network providers, this includes "apps" that can be installed on cell phones. For a cloud provider, the application can be a website or a software product that customers create and host or consume on the cloud.

Getting applications running quickly increases their revenue-generating capability. In addition, consumers are demanding. Enterprises may insist on new features in a consumed application such as their ERP or CRM system; gamers lose interest unless new features are added to Candy Crush or World of Warcraft.

Software developers create applications but may not understand the details of the processes the operations organization must accomplish to integrate the applications into the physical infrastructure.

What often occurs is developers create their application and test it using a limited number of tools. When the application is handed off to the operations group, it may not run in the existing infrastructure environment and must be sent back to the developers to fix the problems delaying, sometimes extensively, getting the application installed, operational, and generating revenue.

Therefore, the "container" was developed to eliminate the integration problem between developers and operations. In the simplest terms, a container is a unit that contains the application code, runtime environment, system tools, operating systems, and libraries necessary to integrate the application into any computing environment (Trend, 2022). Containers are "lightweight," or efficient, and run quickly. Containers streamline development and operations processes and make applications more quickly available and manageable.

14.3.4.1   Hypervisors vs. Containers

You are probably thinking containers sound a lot like hypervisors, and you are correct. However, there are distinct differences (Fig. 14.4).

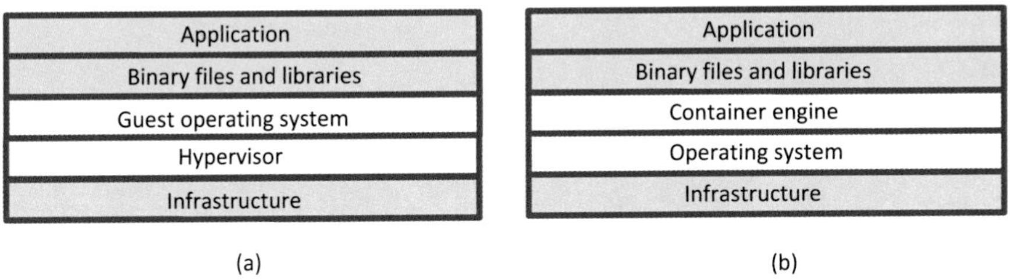

Application
Binary files and libraries
Guest operating system
Hypervisor
Infrastructure

(a)

Application
Binary files and libraries
Container engine
Operating system
Infrastructure

(b)

Fig. 14.4 (**a**) VM vs. (**b**) container

You may recall, a hypervisor virtualizes the hardware of a server and allocates server CPU, memory, and storage resources to a virtual machine (VM). Recall that a Type 1 hypervisor does not have a server host operating system (called a "bare metal" server), but it assigns a scaled-down "guest" operating system to each VM. Each VM can have a different operating system.

Hypervisors have several advantages including allowing multiple operating systems to run on the same server (on the VMs), more efficient use of server resources, and faster service provisioning. However, a VM can be quite large.

On the other hand, containers still require a server (host) operating system and are, therefore, NOT bare metal servers. Since containers share the underlying host operating system instead of each VM having an individual OS, containers do not need to boot an OS or load libraries. Therefore, since containers are "lightweight" they can load in seconds rather than the minutes required by a VM. Containers can also be scaled up and down very quickly. Finally, many more containers can reside on a server than VMs.

There is speculation that containers may replace hypervisors. However, as demonstrated by the diagram above, the hypervisor operates on the south side of the server, interfacing with and virtualizing the server hardware. In contrast, containers operate on the north side of the server, virtualize a host operating system, and are intended to streamline application integration into the server. Therefore, at least for the foreseeable future, hypervisors and containers are complementary cloud technologies.

Containers and their management is now and will become more of a cloud industry focus. Various services that accomplish that are examined next.

14.3.4.2 Docker

The most popular container engine is "Docker." Introduced in 2013, Docker is a Platform-as-a-Service (PaaS) that uses operating system virtualization to deliver software in "containers." A container packages software and all dependencies including libraries and configuration files needed to deploy an application. A Docker container can run on Windows, Linux, and MacOS.

14.3.4.3 Kubernetes

An enterprise as large as Google runs millions of containers at the same time. Managing them is a challenge. Therefore, Google invented Kubernetes, an open-source system for deploying, managing, and scaling containerized applications.

14.3.4.4 AWS Container Services

AWS integrates Docker and Kubernetes using several services, as presented below. Comparable services from Microsoft Azure and GCP are cross-referenced as well.

AWS Elastic Container Service (ECS)

Docker containers can be implemented in AWS using the Elastic Container Service (ECS). ECS and similar services offered by other providers are referred to as "Containers-as-a-Service (CaaS)."

Docker operates using clusters and tasks. In AWS a Docker cluster is a set of EC2 virtual machines running Docker which are acted on by Docker tasks, or instructions. ECS will request an EC2 to create Docker containers. However, you are still managing infrastructure because the configuration of the EC2s must still be defined by you as well as autoscaling and load balancing for each container.

Use cases for ECS includes:

- If your organization is deeply embedded in Docker.
- You still want management control of your EC2s.
- If your existing VPC's and subnets and EC2 autoscaling requirements have already been defined then ECS is a good choice.
- If you run a lot of batch or scheduled jobs that have a duration of 15 minutes or more.

Microsoft Azure: Azure Container Instances
Google Cloud Platform: Container Engine

AWS ECS for Kubernetes (EKS)

ECS for Kubernetes allows Kubernetes to operate within the AWS environment without the operations responsibility of Kubernetes. Once EKS is configured, the management, deployment, and monitoring of your infrastructure will replace AWS processes with those of open-source Kubernetes.

Use cases for EKS include if you want to use:

- AWS but want the option to use other cloud providers.
- The most future-proof container service.
- Open source tools wherever possible.

Microsoft Azure: Azure Kubernetes Service
Google Cloud Platform: Kubernetes Engine

AWS ECS Fargate

Like ECS, AWS Fargate is also part of the next generation of AWS computing. However, Fargate is serverless. If desired, Fargate abstracts operational functions so the customer can focus solely on application development. The customer will not have to configure Docker containers or tasks or any EC2 parameters including load balancing and autoscaling. However, if the customer desires control of EC2s they can be accessed from Fargate.

ECS Fargate is a good choice if:

- Your infrastructure is primarily AWS and you do not have plans to use other providers or non-Docker containers.
- Managing EC2s or containers, which gets more complex with scale, is not desired.
- Traffic fluctuates dramatically, say from day to night.
- Demands are periodic and occasional.

Microsoft Azure: Azure Container Instances
Google Cloud Platform: Cloud Run

14.3.5   DevOps

As previously described, the relationship between application developers and operations that integrates applications into the infrastructure is often disconnected. Developers test their applications with a small set of tools that do not necessarily reflect the actual environment they are expected to run on. One effort to minimize the problem is the container. However, a broader and more defined *processes* needed to be developed to streamline getting applications online.

"DevOps" is a conjunction of "**Development**" and "**Operations.**" It is a framework for creating not only integrated processes but an attempt to also bring people into processes to create a more inclusive development culture. Roles that are traditionally "silos," such as IT, operations, security, quality assurance, testing, and others which often operate as isolated functions, can now, by applying DevOps, work more closely together. DevOps integrates numerous different roles (Fig. 14.5).

Advantages of DevOps include:

- Faster time to market
- Reduced cost
- Fewer failures
- Faster resolution of problems

Fig. 14.5 DevOps roles

Quality
Assurance

DevOps
Engineer

Release
Coordinator

Security
Specialist

Cloud
Architect

Software
Developer

- Improved morale of teams
- Improved communication and collaboration
- Measurability
- Accountability
- More rapid innovation
- Less "firefighting"

DevOps, as the acronym implies, is focused on development and operations. It includes people and processes as well as the development of a unified organizational culture with the goal of a streamlined product pipeline that better serves customer needs.

14.3.5.1   DevOps Application Lifecycle Phases

DevOps defines four phases:

- **Plan.** In the Plan phase the application high- and low-level requirements are developed.
- **Code.** Developers begin writing code based on requirements defined in Plan.
- **Build.** Automation tools are used to build the release code artifacts.
- **Test.** The new build is run through a series of tests to verify it is production ready.
- **Release.** All tests have been passed and build is prepared for production deployment.
- **Deploy.** The build is deployed to production.

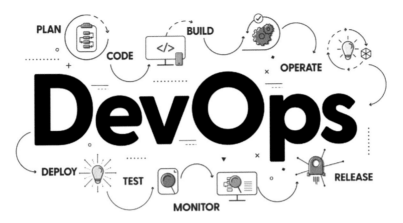

Fig. 14.6 DevOps lifecycle

- **Operate.** The operations team is responsible for keeping the application online and growing as needed.
- **Monitor.** Analytical data is gathered and evaluated. Findings are communicated to enable continuous improvement.

14.3.5.2   DevOps Tools

DevOps is not the first attempt to consolidate different skillsets into a more effective organization. The difference is DevOps defines not only processes but actual tools that are used in each step of the process (Fig. 14.6).

As a cloud engineer you will be working more with DevOps and the tools used to accomplish DevOps processes.

14.4   Evolution of Network Architectures

Cloud infrastructures, such as AWS, are located in data centers remote from customer on-premises facilities. Moving to the cloud has many advantages such as not having to buy, operate, and maintain hardware. However, remote data centers do have one disadvantage and that is delay. It is obvious that having to retrieve data from a location further away will take longer than if it is resident locally.

Although it is not widely advertised, many organizations have been unpleasantly surprised by the slower response times of a cloud-based infrastructure above that of their previously on-premises-based applications. Therefore, cloud providers are implementing ways to decrease application response times.

The following are discussed in more detail:

- Content Deliver Networks (CDN's)
- Caching

- The "edge"
- Internet of Things (IoT)
- SD-WAN

14.4.1 Content Delivery Networks (CDN's)

A solution that many cloud providers have used to decrease delay is the Content Delivery Network (CDN) (Fig. 14.7). When a customer accesses an application via the Internet the request is pulled off at the ISP and routed across the separate network that provides better service. This bypass network is a CDN.

A Content Delivery Network (CDN) is a globally distributed network of web servers or Points of Presence (PoP) whose purpose is to provide faster content delivery. A cloud provider can implement their own CDN or lease capacity from companies that specialize in providing CDN service. Therefore, the CDN can eliminate delays imposed by the public Internet.

The CDN developed and offered by AWS is called CloudFront.

14.4.2 Caching

An addition to the CDN is caching capability. Caching is basically storing data. A good solution is the CDN delivers the content and the caching function stores it close to the customer where it can be quickly accessed. A good example of a CDN using caching is Netflix.

Netflix is one of AWS's largest customers. AWS has implemented caching locations close to large clusters of customers. Netflix then downloads the most popular movies across the AWS CDN using CloudFront to the caches. As a result, the download time for most Netflix movies today occurs very quickly.

Fig. 14.7 Content
Delivery Network (CDN)

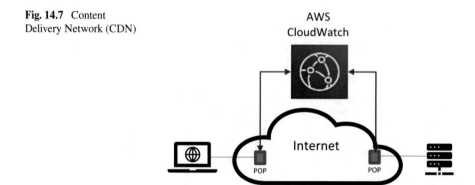

14.4.3   The "Edge"

Caching and CDN accomplish moving data closer to the customer to provide faster and more reliable service. The next evolution is called the "edge." It is the process of moving cloud functions, such as compute, physically closer to the customer where their operation is not just stored but the work accomplished. In other words, instead of cloud functions being centralized in large data centers, they are distributed to hardware on the customer's location.

This transition from centralized to distributed architectures has occurred many times in the past. One case is the introduction of the personal computer where computing was distributed rather than accomplished in a centralized facility by mainframe computers. It occurred again when servers were installed in data centers again centralizing computing resources. As a result, massive "mega" data centers we see today will begin to shrink in size as physical devices and logical services will move more and more to the edge (closer to the customer).

Edge capabilities are in their infancy. Equipment manufacturers are now beginning to offer hardware "micro" data centers. AWS has also recently released their *Outposts* hardware product that provides compute, storage, and database functions close to or on customer premises (AWS, 2023a, b, c).

14.4.4   Internet of Things (IoT)

The Internet of Things (IoT) is the movement that enables devices of all kinds to access the Internet. For instance, smart homes can enable any appliance or home function to be connected to the Internet. As a result you could turn lights in your house on and off, adjust your thermostat, or other functions from your cell phone. However, home functions are not the primary IoT application. An example might be industrial applications that use sensors that gather and distribute discovered data are becoming more common.

IoT will generate massive amounts of data which will need to be compiled and analyzed. Sending all that data from its source all the way to the cloud data center will burden the network and result in data backlogs and delays. Performing processing at the data source and forwarding only results to the data center is more efficient as well as cost-effective. Therefore, IoT is a major driver for the advancement of the edge.

14.4.5   SD-WAN

Traditionally, remote users would access applications by connecting via the Internet to an enterprise data center using an encrypted Virtual Private Network (VPN). If the user wanted to connect to another platform, for instance to AWS, the enterprise data center would create another VPN to connect to AWS.

Fig. 14.8 SD-WAN

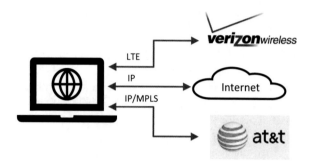

Other WAN technologies were developed that provided alternatives to Internet-based VPNs. Private MPLS networks became available. These connections were secure and more reliable than VPNs. They were more expensive but provided much better service to customers.

Managing many different network connection technologies as well as what applications using each one was difficult. Software Defined Network (SDN) technology promised better solution. Specifically, the Software Defined Wide Area Network (SD-WAN).

SD-WAN is a Network-as-a-Service (NaaS) technology (Fig. 14.8). The way SD-WAN works is it dynamically looks at the data being sent to and from a user and selects the appropriate network connection based on cost requirements or other metrics. A packet to Facebook would be routed over the open Internet while a connection to an AWS account might use an MPLS connection. SD-WAN enable better network connectivity management and provides for more cost-efficient operation (Fig. 14.8).

14.5   Evolution of Cloud Security

Network and cloud security are challenging. New methods must be developed to combat constant cyber-attacks. One method is the Secure Access Service Edge (SASE).

14.5.1   Secure Access Service Edge

Secure Access Service Edge (SASE), pronounced "sassy," is a cloud-based architecture that converges an organization's network and security services (PRISMA, 2023). SASE combines network security functions with SD-WAN. The idea is that SASE targets networking and security shortcomings that traditional WAN and

SD-WAN architectures cannot fully address. While SD-WAN solves many connectivity issues, it doesn't account for the fact that enterprise architectures now focus less on accessing applications in the enterprise data center and more on those in the cloud.

And for security, SD-WAN users must still backhaul traffic to the data center for inspection and authorization. Direct connectivity to outside services is not allowed. This it is inefficient and requires a lot of steps to accomplish in the enterprise data center. What is needed is security based on user authentication and on what service are needed where the user's traffic enters the network. This would eliminate the backhaul through the enterprise data center for security and allow user's direct remote connectivity to cloud and other outside services.

SASE brings security inspection engines closer to the traffic entry points (again, functionality is moved to the edge of the network rather than in the cloud provider's data center) after which traffic can then be forwarded to the Internet or other SASE clients. Ultimately, SASE it converges separate network technologies and network security capabilities. Therefore, SASE is referred to as **Network Security as-a-Service (NSaaS)**.

SASE security is implemented on top of SD-WAN by implementing security inspection and policy enforcement at the network edge and by identifying users based on the context of the connection source, user device, and location.

14.6   Automation and Programmability

Large data centers might install hundreds of hardware devices and virtual machines daily. The scale of the data center operations could not be possible if each device had to be manually configured. Therefore, automation methods were developed that streamlined and accelerated device deployment.

Cloud engineers will be responsible for using many different technologies that automate the configuration and deployment of devices and cloud services. In addition to specific certifications on cloud provider platforms cloud engineers should also gain competency in the technologies below.

14.6.1   Python

In just a few short years the Python programming language has become a standard skillset for networking and cloud operations. Cloud engineers are expected to be able to use basic Python scripts.

14.6.2   RESTful API's

Representational **S**tate **T**ransfer (RESTFUL) Application Programmer Interfaces (API's) are used in the cloud to communicate between cloud services using services. Cloud engineers are increasingly being required to be skilled in the use of RESTful API's.

14.6.3   Ansible and Others Automation Tools

Initially automation was accomplished using scripts written using scripting languages such as PERL. More efficient and specialized methods and products have been developed and implemented.

One product is Ansible which is an open source provisioning and deployment system. Similar products are Chef, Puppet, and Salt. Cloud engineers should begin to familiarize themselves with these products. They will be used more in the near future.

14.6.4   Java Script Object Notation (JSON)

JSON is used to create cloud policies and other functions. Writing JSON policies from scratch is not generally required since cloud providers have hundreds of pre-written JSON policies. However, cloud engineers should be able to modify JSON scripts by adding specific attributes is often necessary.

14.6.5   Yet Another Markup Language (YAML)

YAML is a data serialization language used to write configuration files. This language will be used more and more in the future as network and cloud automation methods advance.

14.7   Summary

This section covered many topics. Cloud services management examined the differences between managed and unmanaged services. An example of an unmanaged service is AWS EC2; a managed service example is AWS RDS. Applications development has evolved from monolithic approaches to applying Service-Oriented

Architectures, microservices, and containers. Container technologies include Docker and Kubinetes. Serverless technologies such as AWS Lambda automate the use of web applications. The lifecycle of web applications are streamlines using DevOps principles and tools. SD-WAN Network architectures changes discussed are SD-WAN, Content Delivery Networks (CDN's), caching, the "Edge," and Internet of Things (IoT), network architectures. Network and cloud security is evolving by using Secure Access Service Edge (SASE). Finally, automation and programmability technologies that cloud engineers should be familiar with including Python, Ansible, RESTful API's, and YAML are introduced.

Homework Problems and Questions

1. How are managed and unmanaged services different? Give an example of each.
2. Compare and contrast the use Service-Oriented Architectures (SOA) microservices.
3. Discuss the advantages of using containers.
4. Describe how DevOps processes and tools improve web applications development.
5. Describe the use and advantages of SD-WAN.
6. How does SASE enhance cloud security?
7. Compare the operation and use of caching and the "edge."
8. Describe tools used for cloud automation and programmability.

References

AWS. (2023a). *AWS outposts family.* [Online] Available at: https://aws.amazon.com/outposts/
AWS. (2023b). *Build a serverless web application.* [Online] Available at: https://aws.amazon.com/getting-started/hands-on/build-serverless-web-app-lambda-apigateway-s3-dynamodb-cognito/
AWS. (2023c). *Choosing between Amazon EC2 and Amazon RDS.* [Online] Available at: https://docs.aws.amazon.com/prescriptive-guidance/latest/migration-sql-server/comparison.html
Gill, N. S. (2022). *Service-oriented architecture (SOA) | A quick guide.* [Online] Available at: https://www.xenonstack.com/insights/service-oriented-architecture
PRISMA. (2023). *What is SASE?.* [Online] Available at: https://www.paloaltonetworks.com/cyberpedia/what-is-sase
Redhat. (2020). *What is Function-as-a-Servcie (FaaS)?.* [Online] Available at: https://www.redhat.com/en/topics/cloud-native-apps/whatis-faas
The Upwork Team. (2021). *Visualizing the DevOps team structure: Roles and responsibilities.* [Online] Available at: https://www.upwork.com/resources/visualizing-devops-team-structure-roles-responsibilities
Trend. (2022). *What is a container?.* [Online] Available at: https://www.trendmicro.com/vinfo/us/security/definition/container

Index